AGRICULTURE IN UGANDA

Volume III

FORESTRY

Editor

Joseph K. Mukiibi

Fountain Publishers / CTA / NARO

Published by:
Fountain Publishers Ltd
P.O. Box 488
Kampala
Uganda

In cooperation with:
National Agricultural Research Organisation (NARO); and
The ACP-EU Technical Centre for Agricultural and Rural Cooperation (CTA).
The Technical Centre for Agricultural and Rural Cooperation (CTA) was established in 1983
under the Lome Convention between the ACP (African, Caribbean and Pacific) Group of
States and the European Union Member States.

CTA's tasks are to develop and provide services that improve access to information for
agricultural and rural development, and to strengthen the capacity of ACP countries to produce,
acquire, exchange and utilise information in this area. CTA's programmes are organised around
four principal themes: developing information management and partnership strategies needed
for policy formulation and implementation; promoting contact and exchange of experience;
providing ACP partners with information on demand; and strengthening their information and
communication capacities.

Postbus 380
6700 AJ Wageningen
The Netherlands

ISBN 9970-02-240-7

Other volumes of *Agriculture in Uganda* include:
   Volume 1: General Introduction ISBN 9970-02-243-1;
   Volume II: Crops ISBN 9970-02-234-2; and
   Volume IV: Livestock and Fisheries ISBN 9970-02-239-3

Editorial Board

Contents

List of abbreviations

NEAP: National Environmental Action Plan

FAO: Food and Agriculture Organisation of the United Nations

UNCED: United Nations Conference for Environment and Development

UNESCO: United Nations Educational, Scientific and Cultural Organisation

MAB: Man and the Biosphere

PAR: Photosynthetically Active Radiation

MONR: Ministry of Natural Resources

MPED: Ministry of Planning and Economic Development

NORAD: Norwegian Agency for Development Cooperation

SAARI: Serere Agricultural and Animal Production Research Institute

LIRI: Livestock Health Research Institute

NAARI: Namulonge Agricultural & Animal Production Research Institute

List of contributors

Byabashaija, D. M., Research Officer, FORI

Esegu, J.F.O., Senior Research Officer, FORI

Kahembwe, F., Research Officer, FORI

Kiwuso, P., Research Officer, FORI

Maiteki, G.A., Research Officer, FORI

Mbalule, M.I., Research Officer, Agroforestry Programme, FORI

Ndemere, P., Research Officer, FORI

Turyareeba, P.J., Research Officer, FORI

Wajja-Musukwe, N., Senior Research Officer, FORI

Acknowledgements

This book has been published as part of the commemorations to mark 100 years of agricultural research in Uganda. I was inspired by the need to update Jameson's 1970 *Agriculture in Uganda* which had become completely outdated. When I discussed this matter with my colleagues, I was encouraged to proceed and to set up a Task Force to handle the matter, and coordinate the production of the book. I am very grateful to my colleges for this encouragement.

I gratefully acknowledge these members for their patience and perseverance in the seemingly endless, yet successful, task of producing this book. The Task Force engaged a number of authors who worked tirelessly to produce the various chapters of the book. I wish to acknowledge the contribution of each of the authors. The authors visited various personalities, both in government and the private sector, institutions and organisations, to solicit for relevant information, for which I am extremely grateful. I also wish to acknowledge the reviewers, who provided valuable and thorough critiques of the various manuscripts. Because of limitation of space, I cannot name them here.

Dr Peter Esele, then Director of Research, Serere Agricultural and Animal Production Research Institute (SAARI), coordinated the production of this book. I am grateful for his contribution and tireless efforts without which the production of this book would have been difficult.

The maps in this book are copied from the National Biomass Study, Forest Department and from Uganda Agriculture, a World Bank country study. I wish to acknowledge them as sources for valuable information.

I also would like to appreciate the guidance and direction provided to me by the chairman, Prof. F.I.B. Kayanja, and members of NARO Board. Their guidance and direction has been very instrumental in shaping NARO into what it is now. Without a strong NARO, the production of this book would have been impossible.

This book has been published with support from CTA and the Government of Uganda. On behalf of the National Agricultural Research Organisation, I would like to express our deep appreciation to CTA and the Government of Uganda for the support. Further, allow me to thank the Managing Director and staff of Fountain Publishers Ltd for the great job they have done in turning the manuscripts into a book of this nature. It is not an easy job going through thousands of pages of manuscript of this type. Fountain Publishers, I salute you! Apart from serving the current generation, this book will be a useful reference to future generations on Uganda's agriculture during the period under review. I commend this book to current and future generations.

J. K. Mukiibi
Director General, NARO
and *Editor-in-Chief*

Foreword

Agriculture is the mainstay of the African economy where it contributes between 29 - 58% GDP, 66 - 100% of export earnings and employs 68 - 92% of the population. It is therefore the main source of income, employment, food, foreign exchange; and it supplies raw materials for domestic industries. Agricultural growth is essential for economic growth and poverty eradication in the continent and very few countries have had rapid growth rates without agricultural growth preceding or accompanying it.

There is a considerable amount of emphasis being given to the concept of sustainable agriculture. In Uganda, this is termed modernisation of agriculture. The country has a total land area of 241, 038 km², 81% of which is suitable for agriculture. With a favourable climate, it has some of the best agricultural land in Africa. The soils and climate over the greater part of the country are favourable and permit farming throughout the year. Consequently, the country is able to produce a wide variety of tropical and sub- tropical agricultural products all year round.

The agricultural sector in Uganda for the foreseeable future, will remain the mainstay and engine of growth for the economy. Over 88% of the country's population live in the rural areas and earn their livelihood from agriculture. In1996/97, the sector contributed 47% of the total GDP and over 94% of exports. The contribution of agriculture to GDP has been expanding each year with an average growth rate of 6.4% per annum (1988 to 1995). The country continues to depend mainly on the agriculture sector for its foreign exchange earnings, and like many other developing countries Uganda has mainly agro-based industries, which depend on the sector for provision of raw materials.

Agricultural production in Uganda can be categorized in four major sub-sectors, namely crops, livestock, fisheries and forestry. The crops sub-sector is by far the largest in terms of area of coverage and contribution to GDP (76% in 1995). Much of the food crop production, however, is for direct consumption at the household level. Only 33% is sold to the domestic and export markets. Smallholder farming, estimated at 2.5 to 3.0 million holdings, predominates, accounting for 94% of all crop production. The typical smallholder farmer cultivates 0.5-2.0 ha per year. Plantation farming, which takes 6% of the cultivable area, consists entirely of sugar cane and tea estates. They are now being run by the private sector as companies and outgrowers.

The livestock subsector accounts for 17% of the total agricultural GDP and 10% of the total GDP. About 66% of the livestock products reach the market, compared to only 33% of the food crop output. In Uganda, farm animals of commercial importance include cattle, goats, sheep, pigs and poultry. Livestock is an integral part of the farming systems in Uganda and nearly 30% of the farming households depend on livestock for the major part of their incomes. The production potential is high because quality pastures can be grown all the year round, and there is an increase in grain milling capacity and livestock feed availability. Commercial mixed farms, smallholders and pastoralists constitute the national cattle herd. Ninety percent of this herd is composed of indigenous breeds. Pastoralists are found mostly in the north-east (Moroto and Kotido districts) and southwest (parts of Mbarara, Ntungamo, Bushenyi, Sembabule and Masaka districts) of the country. Commercial dairy farming is based on imported dairy breeds, (mainly friesian) and crosses with indigenous cattle. Commercial dairy farming is developed primarily in the fertile cresent around the northern and western shores of Lake Victoria and in the south western region. Southwestern Uganda now produces about 90% of the marketed milk. A commercial cattle ranching sector was established in the 1950s. By 1984, ranches were producing about 20% of Uganda's beef. Commercial poultry and pig production has been growing steadily, based, almost entirely, on small units.

In the livestock sector, the major animal diseases are rinderpest, contagious bovine pleuropneumonia (CBPP), trypanosomiasis, foot and mouth disease, tickborne diseases, rabies, Africa swine fever and Newcastle disease.

The Fisheries sub-sector accounted for 5% of agricultural GDP in 1996/97. Almost the entire production is monetised with the domestic market consuming 85% of the total catch. Processed fish for exports make up the remaining 15%. In 1996, this sector was second to coffee in export earnings estimated at US $ 40 million. Up to 17% of Uganda's surface area is covered by lakes, rivers and swamps (about 42,000 km^2). About 11% of Uganda's animal protein is from fish. The fish industry provides the cheapest source of animal protein for Ugandans. The demand for fish is constantly growing: yet most of the traditional sources of fish are already being fully exploited. The country is therefore putting emphasis on management, surveillance and control measures to conserve fish species, breeding stocks and fishing practices to achieve targeted sustainable yields. Aquaculture development has received priority attention in order to meet the ever increasing demand for fish.

The forestry sub-sector, comprising 14,900 km^2 of natural forests and savanna woodland on gazetted reserves and about 20,000 ha of plantation forests, presently accounts for about 4% of Uganda's agricultural GDP. The net output stands at 54,000 m^3 of sawnwood, one million poles, 18 million m^3 of fuelwood and 100,000 mt of charcoal. Although, at present, the contribution of forestry to foreign exchange is little, it has tremendous potential for import substitution, export earnings, economic wellbeing and environmental security. Natural and plantation forests are widely distributed throughout the country. Many areas also feature agro-forestry activities. Moreover, there are important trends in the development of social and community forestry.

A major challenge to the agricultural sector is to develop and adopt the technologies which can increase the overall on-farm production and productivity. This increase can only occur through increased utilisation of modern technological inputs (e.g. high yielding seed varieties, fertilizers, mechanisation, water management, improved livestock breeds, etc). In order to maintain sustainability of the natural resource base, technologies that ensure minimum environmental degradation (minimum soil erosion, development of alternative sources of energy to fuelwood, agroforestry, etc) are also available. The National Agricultural Research Organisation ensures the efficient supply and utilisation of these technologies in timely and affordable manner.

Government has adopted a policy to modernize agriculture. This is often referred to as the Plan for the Modernisation of Agriculture *(PMA)*. The PMA aims to achieve a transformation of the currently predominant subsistence farming into a dynamic and profitable commercial agriculture where farmers produce for the market. The transformation of the agriculture sector will require readily available knowledge and technologies for users. The publication of *Agriculture in Uganda* by NARO is a great stride in the right direction.

Agriculture in Uganda gives a comprehensive account of the agriculture in this country. The first edition was by J. Tothill and was published by Oxford University Press in 1940. Tothill's book was updated by J.D. Jameson and also published by Oxford University Press in

1970. It provides an up-to-date account of agriculture in the country for the benefit of all stakeholders in the sector. It was also intended to contribute to the study of agriculture in the tropics.

It is almost three decades since the last edition of the book by Jameson appeared in 1970. Since then, the agriculture sector in Uganda has undergone a number of developments, taking advantage of the economic and technical changes sweeping across the nation and the world at large. Consequently a new edition of the book was felt necessary. It is in this context that the National Agricultural Research Organisation (NARO) undertook to write a new edition which is completely a new book. NARO has published a radically new book which focuses on many of the strategic issues facing Uganda's agriculture. The book is intended to provide guidelines to farmers, non-governmental organisations, agro-based industrialists, researchers, extension workers and civic leaders, that will ensure increased and sustainable agricultural production and productivity. It will also be useful for teachers and students of agriculture in schools, colleges and universities. The book will also be of interest to the rest of Africa and other collaborating partners in agricultural research and development in the tropics since agriculture is the mainstay of the economies of the countries in this region. Specifically it gives a detailed account of the relevant issues that would contribute to the country's drive towards modernization of agriculture and will guide the planning and execution of agricultural development programmes in the country. I recommend it to every one who is interested in the development of agriculture in this country and the tropics at large.

In this book, agriculture has been used in the context of its broad description by the Food and Agriculture Organisation (FAO) of the United Nations. It embraces crops, forestry, fisheries, livestock, the environment and related disciplines. There has not been a comprehensive write-up of the livestock, fisheries and forestry sub-sectors in Uganda. This is the first detailed account of these sub-sectors. Detailed accounts are provided in each of these disciplines, as they relate to the agricultural sector in Uganda.

I wish to thank NARO its scientists and the Ministry of Agriculture, Animal Industry and Fisheries for the great work they have done in producing a book of this nature. It is a sign of NARO's great commitment to the development of this country.

In 1996, the president of Uganda, His Excellency Yoweri Kaguta Museveni, had a vision to modernise agriculture in this country. Since this time, the president has worked tirelessly to ensure that this vision is turned into reality. I am happy to note that in December 2000, the Plan for the Modernisation of Agriculture (PMA) was launched, the implementation of which has now started. As already pointed out, this book is part and parcel of the effort to modernise agriculture. Further, the President has been in the forefront of the drive to spread modern farming . He received a national award for the improvement of the Ankole cow in 1998 and an international prize for the sustainable end to hunger in Africa awarded by the Hunger Project. Because of these and his inspirational leadership, I wish to appreciate the personal efforts of president Museveni in the drive to modernise agriculture in the country. His vision inspired NARO to produce this book.

Dr Speciosa Wandira Kazibwe *(MP)*
Vice President of the Republic of Uganda
Kampala, May 2001

1

FOREST TREE GENETIC RESOURCES IN UGANDA

J. F. O. Esegu

Uganda is a landlocked country covering about 236,000km^2 in the central African plateau. Ecologically, the country is diverse due largely to its location in the east-central Africa where there is an overlap between ecological communities characteristic of the dry East African savannah and those of the West African rainforests. This is also coupled with the country's great topographical diversity which ranges from about 600m above sea level at the bottom of the Rift Valley to over 5,000m above sea level at the peak of the Rwenzori Mountains.

The UNESCO vegetation map of Africa (White, 1983), shows that seven out of the eighteen Africa's mainland phytochoria or biogeographical regions are represented in Uganda. The highly diverse landscape features: rift valleys, highlands, mountain ranges, papyrus swamps, *Acacia* savannahs, and an extensive network of rivers and lakes. These constitute the largest number in any single country (Howard 1991).

The country's vegetation and associated animal communities are characteristic of habitats as diverse as the glacier topped mountains, lowland rainforests, woodland savannah and deciduous bush land and thickets (White 1983). Uganda's vegetation communities have been described in the chapter on vegetation.

Most of Uganda's biodiversity is in the natural forests, wetlands, open waters, and dry and moist savannah. It is estimated that the country has about 250,000 to 500,000 species altogether, with flowering plants numbering 4,500 species (Howard 1991). A list of tree species from twelve of Uganda's principal forest reserves (Howard 1991) is in Appendix 1.1 (p.6). The summary of the number of tree species in the twelve principal forest reserves is in Table 1.1 and the exotic trees, compiled from Forestry Research Records (1998) is in Appendix 1.2 (P.21).

Current forest tree genetic resources issues

Uganda, usually referred to as The Pearl of Africa has experienced severe degradation of her natural resources. According to the Uganda National Environmental Action Plan (NEAP) and FAO, the country's annual rate of deforestation is 500km^2 and 650km^2, respectively (NEAP 1992). The main cause of deforestation in Uganda is uncontrolled harvesting of timber, wood fuels, non-timber products and encroachment of cultivation of agricultural crops into the forests due to population pressures.

Table 1.1 Summary of Number of Tree Species in the Uganda Principal Forest Reserves

	Forest	Species Recorded in Literature / Herbarium	Species Recorded during the Howard (1991) Study	Total
1.	Kibale	199	6	205
2.	Samliki	75	92	167
3.	Budongo	241	0	241
4.	Kalinzu-Maramagambo	216	21	237
5.	Bugoma	144	11	155
6.	Bwindi	151	4	161
7.	Kasyoha-Kitomi	135	69	204
8.	Itwara	18	131	149
9.	Sango Bay	99	73	172
10.	Mabira	128	70	198
11.	Mount Elgon	111	0	111
12.	Rwenzori	52	21	73

In recognition of the importance of biodiversity in Uganda, the country has ratified the Convention on Biological Diversity in 1993. Despite previous efforts aimed at conserving critical sites through formal protected areas such as forest reserves, national parks, wildlife reserves and the introduction of comprehensive National Environmental Management Policy and Legislation, biodiversity continues to be threatened with extinction. A review of deforestation in Uganda has been done by Hamilton (1984).

Due to this rather worrying development, Uganda, following the recommendations of the United Nations Conference for Environment and Development (UNCED), is developing the institutional capacity and legal framework for forestry and the environment in general. Following the implementation of the Environment Action Plan, Uganda is to adopt a new forestry legislation and to launch the National Forestry Action Plan for the management, conservation and sustainable development of forests which should be integrated with other land uses.

Forest nature reserves (*in situ* conservation)

Forest nature reserves are areas that are designated by the Forest Department to be protected from all forms of extractive resource use. They are *in situ* conservation areas whereby species grow wild in their environments; the legal protection afforded to them is derived from the Forest Act (Government of Uganda, 1964). The objectives of maintaining areas of forest as nature reserves which have been described by Howard (1991) are to:

- ensure the long-term survival of as many as possible of Uganda's forest species;

- maintain viable, representative samples of each of the major forest types in an undisturbed state;

- provide suitable areas for fundamental scientific research, education and other forms of non-consumptive use including recreation and tourism; and

- help sustain the productive capacity of adjacent production forests, and their role in environmental protection by serving as a reservoir of plant and animal species able to disperse and recolonise these areas.

In addition to other forest areas within the national and forest parks the Forest Department has designated 20% of the natural forests as nature reserves. The proportion of each forest reserve has been decided upon considering the number and rarity of the species it supports, the diversity of its habitat, its potential for timber production and non-consumptive forms of use such as research, education, recreation and tourism.

The protected area network is to ensure the preservation of viable representative samples of all major forest communities and guarantee the diversity of ecological interactions contained within them. This is best achieved through establishing several large protected areas in different geographical regions. They will have at least one in each altitudinal zone so as to include the greatest possible range of habitat types, the richest examples of these habitats, and the widest possible altitudinal continuum. The network is supplemented with a large number of smaller subsidiary ones selected to protect additional distinctive habitats and regional habitat sub-types, to protect sites of special importance to particular species, and to provide additional areas for research, education, recreation and tourism development.

In accordance with the principles of reserve design developed under UNESCO's Man and the Biosphere (MAB) Programme (Batisse, 1985), each forest reserve is zoned into concentric resource management areas: the core, the buffer, and high intensity management areas. As a broad guideline the core covers 20% of the central part of the forest; whereby the primary objective is the preservation and maintenance of biotic diversity surrounded by concentric zones in which management becomes progressively more intensive towards the reserve boundary.

Introduced species (an *ex situ* conservation)

In Uganda where climate and soil-water conditions are favourable, the evaluation of a large number of tree species offers unique opportunities for improving forest plantation establishment and development. The introduction and evaluation of exotic tree species in Uganda started at the beginning of this century and was intensified in the 1950s. The list of exotic species compiled from the Forestry Research Records (1998) is in Appendix 1.2 (p. 21). These were introduced in Uganda for timber, fuelwood, and poles production, environment conservation and to some extent the supply of non-wood products such as gums and tannin as well as improving productivity of savannah woodlands.

Extensive provenance evaluations of some introduced species have been carried out in collaboration with the Oxford Forestry Institute, United Kingdom. These studies centred largely on *Eucalyptus* and *Pinus* species for rapid growth and wood production. Results from these studies now help to determine provenances selected for afforestation programmes in many parts of the country. They also form the basis for further improvement research.

As a base for tree improvement, introduced species and their provenances are conserved. This is the *ex situ* conservation and is mainly for reasons of economic value, and fear of extinction that they are established and preserved in plantations of seed stands. These conservation stands aim to protect plant genetic resources and ensure maintenance of seed sources of proven species and provenances. *Ex situ* conservation is required for specific ecologies, and allows for easy utilisation of the genetic resource for development, and for further research. The conservation strategy of introduced plantation tree species together with their provenances are conserved in seed stands and through use in the various suitable planting sites.

Various methods are currently in use in Uganda for *ex situ* conservation particularly of exotic tree species. These include:

• *Arboreta:* This is typically the first stage in introducing exotic species. The arboreta in Uganda were created with tree introductions from various sources at the beginning of this century. Most tree species were represented only by a few trees. Many are with known origin but are of limited use for gene conservation due to very low numbers of individuals in the stands.

- *Botanic Garden:* Uganda has a big botanic garden at Entebbe, The Entebbe Botanical Gardens, covering land area of 40ha and holding a diversity of important living collections both native and introduced totaling 352 species. This was the first agriculture and forestry research unit in Uganda.

- *Species Populations, Evaluation Trials and Improvement Programmes:* These include species and provenance trials of many introduced species located at various sites throughout the country. In addition to conserving species and provenances germplasm, the trials and populations provide some information on the requirements of the species with regard to site characteristics.

References

Batisse (1985). Action Plan for Biosphere Reserves, Environmental Conservation 12(i): 17-27.
Government of Uganda (1964). The Forest Act. Laws of Uganda
Hamilton C (1984). Deforestation in Uganda, Nairobi Oxford University Press
National Environment Action Plan (1992); The State of Environment, Uganda Government
Howard C (1991). Nature Conservation in Uganda's Tropical Forest Reserves. IUCN Tropical Forest Programme; pp 313.
White (1983). The Vegetation of Africa. Natural Resources Research to Paris: UNESCO

List of tree species recorded from twelve of Uganda's principal forest reserves (Howard 1991)

TREE SPECIES	FOREST NUMBER											
	1	2	3	4	5	6	7	8	9	10	11	12
1. Cyathea manniana	+	•	•	+	•	+	+	•	•	•	+	+
2. C.dregei	•	•	•	•	•	•	•	•	•	•	+	•
3. C.camerooniana	•	•	•	•	•	+	•	•	•	•	•	•
4. Podocarpus milanjianus	•	•	•	•	•	+	•	•	+	•	+	+
5. P. gracilior	•	•	•	•	•	+	•	•	+	•	+	•
6.												
7. Juniperus procera	•	•	•	•	•	•	•	•	•	•	+	•
8. Phoenix reclinata	+	÷	+	+	•	•	+	÷	+	÷	•	•
9. Raphia farinifera	•	•	+	+	+	+	÷	•	÷	•	•	•
10. Elaeis guineensis	•	÷	•	+	•	•	•	•	•	•	•	•
11. Pandanus ugandaensis	+	+	+	•	•	•	+	•	•	•	•	•
12. Dracaena steudneri	+	÷	•	+	•	•	+	•	÷	•	•	÷
13. D. afromontana	•	•	•	•	•	•	•	•	•	•	+	+
14. D.fragrans	•	•	+	+	•	•	+	÷	÷	•	•	÷
15. D.laxissima	+	•	+	•	•	+	•	•	+	•	•	•
16. Ensete ventricosum	•	•	•	+	•	•	÷	•	•	•	•	÷
17. Arundinaria alpina	•	•	•	•	•	+	•	•	•	•	+	+
18. Oreobambos buchwaldii	•	•	+	•	•	•	•	•	•	÷	•	•
19. Senecia erici-rosenii	•	•	•	•	•	•	•	•	•	•	•	+
20. S. amblyphyllus	•	•	•	•	•	•	•	•	•	•	+	•
21. S. adnivalis	•	•	•	•	•	•	•	•	•	•	•	+
22. S. elgonensis	•	•	•	•	•	•	•	•	•	•	+	•
23. S. barbatipes	•	•	•	•	•	•	•	•	•	•	+	•
24. Stoebe sp.	•	•	•	•	•	•	•	•	•	•	+	•
25. Erica arborea	•	•	•	•	•	•	•	•	•	•	+	+
26. E. kingaensis	•	•	•	•	•	•	•	•	•	•	•	+
27. Philippia excelsa	•	•	•	•	•	•	•	•	•	•	+	•

Species												
28. P.trimera	•	•	•	•	•	•	•	•	•	•	+	+
29. P.johnstonii	•	•	•	+	•	•	•	•	•	•	•	+
30. P.benguelensis	•	•	•	+	•	•	+	•	•	•	•	+
31. Euphorbia teke	+	+	+	÷	+	•	•	•	÷	•	•	•
32. E.obovallifolia	•	•	•	•	•	•	•	•	•	•	+	•
33. Euphorbia sp.	•	+	•	•	•	•	•	•	•	•	•	•
34. Elaeophorbia sp.nov.	+	+	•	+	•	•	•	•	•	•	•	•
35. Aningeria altissima	+	÷	+	+	+	•	+	÷	÷	•	•	•
36. A.adolfi-friederici	•	•	•	•	•	•	•	•	•	•	+	+
37. Bequaertiodendron oblanceolatum	+	•	+	•	+	•	•	•	•	+	•	•
38/39. Chrysophyllum muerense	•	+	+	•	+	•	•	•	•	+	•	•
40. C.albidum	+	•	+	+	+	+	+	÷	+	+	•	•
41. C.perpulchrum	•	•	+	•	+	•	•	•	•	+	•	•
42. C.gorungosanum	+	•	•	+	•	+	+	÷	•	•	+	÷
43. C.delevoyi	•	•	•	•	•	•	•	•	+	÷	•	•
44. C.beguei	•	+	•	•	•	•	•	•	•	•	•	•
45. C.pentagonocarpum	•	•	•	•	•	•	•	+	•	•	•	•
46. C.pruniforme	•	•	+	•	•	+	•	+	•	•	•	•
47. Mimusops bagshawei	+	•	+	+	+	•	÷	÷	÷	+	+	•
48. M.kummel	•	•	+	•	•	•	•	•	•	•	+	•
49. Manilkara multinevis	•	•	+	•	•	•	•	•	•	+	•	•
50. M.hutugi	•	•	+	•	•	•	•	•	•	•	+	•
51. M.obovata	•	•	+	•	•	•	•	•	+	•	•	•
52. M.dawei	•	+	+	+	+	+	•	•	•	+	•	•
53. Pchystela brevipes	÷	•	+	÷	+	÷	÷	÷	+	+	•	•
54. P.msolo	•	+	•	•	+	•	•	•	•	•	•	•
55. Afrosersalisia cerasifera	•	•	+	+	•	•	+	÷	•	•	+	•
56. Antiaris toxicaria	+	÷	+	+	+	+	+	÷	÷	÷	•	•
57. Antiaris sp.	•	•	+	•	•	•	•	•	•	•	•	•
58. Mortus tactea	+	÷	+	+	+	•	+	÷	•	+	•	•

59. *Chlorophora excelsa*	+	÷	+	+	+	●	+	÷	●	+	●	●
60. *Treculia african*	+	●	+	+	+	●	÷	÷	÷	●	●	●
61. *Bosqueia phoberos*	●	+	+	●	●	●	●	●	●	●	●	●
62. *Craterogyne kameruniana*	●	+	+	●	●	●	●	●	●	●	●	●
63. *Ficus exasperata*	+	÷	+	÷	÷	●	÷	÷	●	+	●	●
65. *F.vallies choudae*	●	+	+	÷	●	●	+	÷	●	+	●	●
66. *F.mucuso*	+	÷	+	●	÷	●	÷	÷	÷	÷	●	●
67. *F.congensis*	÷	÷	●	÷	●	●	÷	÷	●	÷	●	●
68. *F.urceolaris*	+	÷	+	●	●	●	÷	÷	÷	÷	●	●
69. *F.cafensis*	+	+	+	÷	●	●	÷	÷	÷	÷	●	●
70/71.*F.brachypoda*	+	÷	+	●	●	●	÷	●	÷	÷	●	●
72. *F.brachylepis*	+	÷	+	÷	÷	●	÷	●	●	÷	●	●
73.*F.ingens*	●	●	●	●	●	●	●	●	●	+	●	●
74. *F. polita*	+	●	+	●	●	●	●	●	●	÷	●	●
75. *F.eriobotryoides*	+	÷	+	÷	●	●	÷	÷	÷	÷	●	●
76. *F.vogeliana*	●	+	●	●	●	●	●	●	●	●	●	●
77. *F.cyathistipula*	÷	●	●	÷	●	●	●	●	+	●	●	●
78/79.*F.stipulifera*	÷	●	+	●	●	●	÷	●	+	+	●	●
80/81. *F.pilosula*	●	●	●	●	●	●	●	●	+	÷	●	●
82. *F.natalensis*	+	÷	+	÷	●	+	÷	÷	÷	÷	●	÷
83. *F.thonningii*	●	●	+	●	●	●	÷	÷	●	÷	●	●
84. *F.persicifolia*	●	÷	+	●	●	●	●	●	÷	+	●	●
85. *F.pseudomangifera*	+	÷	+	÷	●	●	÷	÷	÷	÷	●	●
86. *Celtis mildbraedii*	+	+	+	+	+	●	÷	●	●	+	●	●
87. *C.zenkeri*	+	÷	+	+	+	+	●	●	●	+	●	●
88. *C.durandii*	+	÷	+	+	+	●	+	÷	●	+	+	●
89. *C.africana*	+	÷	+	●	+	+	●	÷	●	+	+	●
90. *C.wightii*	●	+	+	÷	+	●	●	●	●	+	●	●
91. *C.adolfi-fridericii*	●	+	●	●	+	●	●	●	●	●	●	●
92. *Trema orientalis*	+	÷	+	+	+	+	+	÷	+	+	+	÷

93. *Holoptlea grandis*	+	+	+	+	+	•	•	•	•	+	•	•
94. *Macaranga schweinfurthii*	÷	÷	+	+	+	+	+	÷	÷	÷	•	•
95. *M.angolensis*	+	•	•	•	•	•	•	•	•	•	•	•
96. *M.monandra*	•	•	•	+	•	+	+	•	+	÷	•	÷
97. *M.pynaertii*	•	•	•	•	•	•	+	•	÷	÷	•	•
98. *M.lancifolia*	•	•	•	÷	•	+	÷	•	÷	•	•	•
99. *M.kilimandscharica*	•	•	•	+	•	+	+	•	•	•	+	+
100. *Neoboutonia macrocalyx*	+	•	•	+	+	+	+	÷	•	÷	+	+
101. *N.melleri*	+	•	+	+	•	•	+	•	÷	•	•	•
102. *Alchornea cordifolia*	•	•	+	+	•	•	+	•	+	÷	•	•
103. *A.laxiflora*	•	+	+	•	+	•	•	•	+	•	+	•
104. *Acalypha neptunica*	+	+	+	+	•	•	•	÷	+	+	+	•
105. *A.ornata*	•	•	+	•	•	•	•	•	•	+	•	•
106. *Croton macrostachyus*	+	÷	+	+	+	+	+	÷	+	÷	+	÷
107. *C.sylvaticus*	+		+	•	•	+	÷	÷	÷	•	+	÷
108. *C.bukobensus*	•	•	•	•	•	+	•	•	•	•	•	•
109. *C.megalocarpus*	+	•	+	+	•	+	+	•	+	+	•	•
110. *Discoglypremna caloneura*	•	•	+	•	•	+	•	•	•	•	•	•
111. *Alangium chinense*	+	÷	+	+	÷	+	+	÷	•	+	+	+
112. *Cordia millenii*	+	+	+	÷	+	•	÷	÷	÷	+	•	•
113. *C.africana*	•	•	+	+	•	•	+	•	•	•	+	•
114. *Ehretia cymosa*	+	•	+	+	+	+	÷	÷	•	+	+	÷
115. *Pterygota mildbraedii*	+	÷	+	+	+	+	+	•	•	•	•	•
116. *Cola gigantea*	+	÷	+	+	+	+	+	•	•	+	•	•
117. *C.bracteata*	+	•	•	+	•	+	•	+	•	•	•	•
118. *Sterculia dawei*	+	+	+	÷	+	•	+	÷	•	+	•	•
119. *Dombeya mukole*	+	+	+	÷	+	÷	÷	÷	•	+	+	•
120. *D.goetzenii*	•	•	•	•	•	+	•	•	•	•	+	+
121. *Leptonychia mildbraedii*	+	•	+	+	•	+	÷	+	•	÷	•	•
122. *Nesogordonia kabingaensis*	•	+	•	•	•	•	•	•	•	•	•	•

123. Strombosia scheffleri	+	•	+	+	+	+	+	÷	•	•	+	•
124. Strombosiopsis tetrandra	•	•	•	•	•	+	•	•	•	•	•	•
125. Heisteria pavifolia	•	•	•	•	•	•	•	•	+	•	•	•
126. Brazzeia longipedicellata	•	•	•	•	•	+	•	•	•	•	•	•
127. Glyphaea brevis	+	+	+	•	+	+	+	•	÷	+	•	•
128. Desplatsia dewevrei	•	+	+	•	+	•	•	•	•	•	•	•
129. D.chyrsochlamys	+	+	+	•	+	•	•	•	•	•	•	•
130. Grewia pubescens	•	+	+	•	•	•	•	•	•	+	•	•
131. G.mildbraedii	•	•	•	•	•	+	•	•	•	•	•	•
132. Caloncoba schweinfurthii	•	÷	+	+	+	•	+	•	•	•	•	•
133. Lindackeria bukobensis	•	•	+	+	•	•	•	•	•	•	•	•
134. L.bequaertii	•	+	•	+	•	•	•	•	•	+	•	•
135. L.mildbraedii	+	+	+	+	+	+	+	•	•	+	•	•
136. L.schweinfurthii	•	+	+	÷	•	•	÷	•	•	+	•	•
137. Dasylepis eggelingii	+	•	+	+	•	+	•	+	•	•	•	•
138. D.racemosa	+	•	•	+	•	+	•	÷	•	•	•	•
139. Rawsonia lucida	+	•	+	+	•	•	+	÷	•	÷	+	•
140. Scolopia rhamnophylla	+	•	•	+	+	•	÷	•	+	÷	•	•
141. Flacourtia indica	•	•	+	•	•	•	+	•	•	÷	+	•
142. Oncoba spinosa	+	•	+	+	+	•	+	•	•	•	+	•
143. O.routledgei	+	•	•	•	•	+	•	÷	•	•	•	•
144. Dovyalis macrocalyx	+	•	+	+	•	•	+	÷	+	÷	+	•
145. D.abyssinica	•	•	+	+	•	•	•	÷	•	•	+	•
146. D.macrocarpa	+	•	•	+	•	•	÷	÷	•	•	•	•
147. Trimeria bakeri	+	•	•	•	÷	•	•	•	•	•	+	•
148. Rinorca ilicifolia	+	÷	+	+	•	•	•	•	÷	+	•	•
149. R.ardisiiflora	+	+	+	÷	+	•	÷	•	•	+	•	•
150. R.brachypetala	+	+	+	•	•	•	÷	÷	÷	•	+	•
151. R.dentata	•	•	+	•	•	+	•	•	+	+	•	•
152. R.oblongifolia	•	•	•	•	•	•	+	•	•	•	+	•

Species												
153. Oruratea densiflora	+	●	+	●	●	●	●	●	+	+	●	●
154. O.hiernii	+	●	+	+	●	+	+	●	+	●	●	●
155. Ochna membranacea	+	●	+	+	+	●	+	÷	●	+	●	●
156. O.bracieosa	●	+	+	+	●	●	●	●	●	+	●	●
157. O.afzelii	●	●	●	●	●	●	●	●	÷	+	●	●
158. O.holstii	+	●	●	+	●	●	●	●	●	●	+	●
159. Hugonia platysepala	●	●	+	+	●	●	+	●	●	●	●	●
160. Cassine aethiopica	+	●	+	+	+	+	+	●	●	●	+	+
161. Maytenus undata	+	●	●	+	●	+	÷	÷	÷	+	●	●
163. M.acuminata	●	●	●	+	●	+	●	●	●	●	+	+
164. Maesa lanceolata	+	÷	+	+	●	+	+	÷	÷	●	●	+
165. Ilex mitis	+	●	●	●	●	+	●	●	+	●	+	+
166. Myrica kandtiana	●	●	●	●	●	●	●	●	÷	●	●	●
167. Aeglopsis eggelingii	+	●	+	●	●	●	+	÷	●	●	●	●
168. Rhamnus prinoides	+	●	●	●	●	●	●	●	●	●	+	●
169. Nidorella arborea	●	●	●	●	●	●	●	●	●	●	+	●
170. Vernonia conferta	+	●	●	●	●	●	÷	÷	●	●	●	●
171. V.sp.aff. V.adolfi-friderici	●	●	●	●	●	●	●	●	●	●	●	+
172. Alchomea floribunda	●	÷	+	+	●	+	÷	●	÷	÷	●	●
173. A.hirtella	+	●	●	+	●	+	+	÷	+	÷	●	+
174. Argomuellera macrophylla	+	+	+	+	●	●	+	●	+	+	●	●
175. Pseudagrostistachys ugandansis	●	●	●	●	●	●	●	●	+	●	●	●
176. Claoxylon hexandrum	●	●	+	●	●	●	●	●	●	+	●	●
177. Maesobotrya purseglovei	●	●	●	●	●	+	●	●	●	●	●	●
178. Securnega virosa	+	÷	+	+	+	●	+	÷	+	÷	+	●
179. Sapium ellipticum	+	÷	+	+	+	+	+	÷	+	+	●	●
180. S.leonardii-crispi	●	●	●	+	●	+	●	●	●	●	●	●
181. Suregada procera	+	÷	+	●	÷	●	●	÷	+	+	●	●
182. Drypetes gerrardii	+	÷	+	+	+	+	+	÷	●	+	+	÷
183. D.Drypetes sp.	●	●	+	+	●	+	●	●	●	●	●	●

184. D.ugandensis	+	•	+	•	•	•	•	•	•	+	•	•
185. D.bipindensis	+	+	+	+	•	+	•	•	•	•	•	•
186. Paropsia guineensis	•	•	•	•	+	•	•	•	•	+	•	•
187. Ficalhoa laurifolia	•	•	•	•	•	+	•	•	•	•	•	•
188. Melchiora schiebenii	•	•	•	•	•	+	•	•	•	•	•	•
189. Maesopsis eminii	•	•	+	+	+	+	+	÷	+	+	•	•
199. Prunus africana	+	•	+	+	+	+	+	÷	÷	÷	+	÷
200. Parinari excelsa	+	•	+	+	+	+	+	÷	+	•	•	•
201. Warburgia ungandensis	+	•	+	+	+	•	+	÷	+	+	•	•
202. Pycnathus angolensis	•	÷	+	+	+	•	+	•	÷	÷	•	•
203. Staudtia kamerunensis	•	•	+	•	+	•	•	•	•	+	•	•
204 Beilschmedia ugandensis	+	÷	÷	+	•	+	+	÷	+	+	•	÷
205. Ocotea usambarensis	•	•	•	+	•	+	•	•	•	•	•	+
206. Ocotea kenyensis	•	•	•	•	•	+	•	•	•	•	•	•
207. Casearia engleri	+	•	+	+	+	+	•	÷	÷	•	•	•
208. C.battiscombei	+	•	•	+	•	•	•	•	•	•	+	•
209. Klainedoxa gabonensis	•	+	+	+	+	•	•	•	+	+	•	•
210. Irvingia gabonensis	•	•	+	•	•	•	•	•	÷	+	•	•
211. Diospyros abyssinica	+	÷	+	+	+	+	+	÷	+	+	+	•
212. Uvariopsis congensis	+	÷	+	+	+	•	÷	÷	•	+	•	•
213.Creenwayodendron suaveolens	•	•	+	•	+	•	•	•	÷	+	•	•
214. Cleistopholis patens	•	÷	+	+	+	•	•	•	•	•	•	•
215. Isolona congolana	+	+	•	•	•	•	•	÷	•	•	•	•
216. Xylopia aethiopica	•	•	•	•	•	•	•	+	+	•	•	•
217. X.staudtii	•	•	•	•	•	+	•	•	•	•	•	•
218. X.parviflora	•	+	•	•	÷	•	•	•	•	•	•	•
219. Monodora myristica	+	•	+	+	+	•	+	÷	•	+	+	•
220. M.angolensis	•	•	+	+	•	•	÷	•	•	•	•	•
221. Uvariodendron magnificum	•	•	•	•	•	•	+	•	•	•	•	•

222. *Uvaria angolensis*	•	•	+	•	•	•	•	÷	•	•	•	•
226. *Turraea floribunda*	+	•	+	•	•	•	•	•	•	+	•	•
227. *T.robusta*	+	•	•	+	+	•	+	•	÷	+	•	•
228. *T.vogelii*	+	+	+	•	•	•	•	•	•	•	•	•
229. *T.vogelioides*	+	÷	+	+	+	•	+	÷	+	+	•	•
230. *Baphia wollastonii*	•	+	+	+	•	•	+	•	•	•	•	•
231. *B. capparidifolia*	•	+	•	•	•	•	•	•	•	•	•	•
232. *Baphiopsis parviflora*	+	÷	•	+	+	•	+	•	+	+	•	•
233. *Maerua duchesnei*	•	•	+	+	•	•	+	•	•	+	•	•
234. *Tapura fischeria*	+	÷	+	+	+	•	•	•	•	+	•	•
235. *Trichocladus ellipticus*	•	•	•	•	•	•	•	•	+	•	+	•
236. *Barteria acuminata*	•	•	•	•	•	•	•	•	+	•	•	•
237. *Apodytes dimidiata*	+	•	•	•	+	•	•	•	+	•	•	•
238. *Leptaudus daphnoides*	+	•	+	•	•	•	÷	•	+	+	•	•
239. *L.holstii*	•	•	•	+	•	•	+	•	+	•	•	•
240. *Uapaca guineensis*	•	•	•	+	•	•	÷	•	÷	•	•	•
241. *Spondianthus preusii*	•	•	+	•	•	•	•	÷	+	•	•	•
242. *Tetrorchidium didymostemon*	•	•	+	+	÷	+	+	÷	+	+	•	•
243. *Bridelia micrantha*	+	÷	+	+	+	+	+	÷	+	÷	+	+
244. *B.brideliifolia*	•	÷	•	+	•	•	+	•	÷	•	•	+
245. *Antidesma laciniatum*	•	•	+	+	•	•	+	•	•	•	•	•
246. *A.membranaceum*	+	÷	+	•	•	+	•	÷	+	+	•	•
247. *Phyllanthus discoideus*	+	÷	+	+	+	+	+	÷	÷	÷	•	•
248. *P.sp.aff.P.inflatus*	•	•	+	+	•	•	+	•	•	+	•	•
249. *Thecacoris lucida*	•	+	+	+	•	•	•	•	•	•	•	•
250. *Cleistanthus polystachyus*	+	•	+	•	•	•	•	•	÷	÷	•	•
251. *Microdesmis puberula*	•	•	+	•	•	•	•	•	•	•	•	•
252. *Chaetacme aristata*	+	÷	+	+	+	•	+	÷	+	+	•	•
253. *Pittosporum mannii*	+	•	•	+	•	+	÷	•	÷	•	•	•
254. *P.spathicalyx*	•	•	•	+	•	+	•	•	•	•	•	•

Species												
255. P.viridiflorum	●	●	●	●	●	●	●	●	●	●	+	●
256. Peddiea fischeri	+	●	●	+	●	+	÷	÷	+	÷	+	+
257. Erythroxylum fisheri	●	●	●	+	●	●	●	●	●	●	●	●
258. Myrica salicifolia	●	●	●	+	●	+	●	●	●	●	●	●
259. Faurea saligna	●	●	●	+	●	+	+	●	●	●	●	+
260. Protea kilimandscharica	●	●	●	●	●	●	●	●	●	●	+	●
261. Agauria salicifolia	●	●	●	+	●	+	●	●	●	●	+	●
262. Rapanea rhododendroides	●	●	●	●	●	+	●	●	●	●	+	+
263. Euclea latideus	●	●	●	●	●	●	●	●	●	●	●	●
264. Nuxia congesta	●	●	+	●	●	+	●	●	●	●	+	●
266. Premna angolensis	+	÷	+	+	+	+	+	÷	●	+	●	●
267. Alstonia boonei	●	÷	+	+	+	●	●	●	÷	+	●	●
268. Rauvolfia oxyphylla	+	÷	●	+	+	●	+	●	●	+	+	●
269. R.vomitoria	●	÷	+	+	●	●	+	●	÷	+	●	●
270. Pleiocarpa pycnantha	+	●	●	+	●	+	÷	+	+	÷	●	●
271. Funtumia africana	+	●	+	+	+	+	+	+	÷	+	●	●
272. F. elastica	●	+	+	●	+	+	●	●	●	+	●	●
273. Tabernaemontana holstii	+	●	+	+	+	+	+	÷	÷	+	●	+
274. T.johnstonii	+	●	●	●	●	●	●	●	●	●	+	●
275. T.usambarensis	+	+	●	+	÷	●	+	●	●	●	●	●
276. T.odoratissima	+	●	●	+	●	+	●	÷	●	●	●	●
277. Picralima nitida	+	+	+	●	●	●	●	●	●	+	●	●
278. Voacanga thouarsii	÷	+	●	+	●	●	+	÷	÷	●	●	+
279. Symphonia globulifera	+	●	●	+	●	+	+	÷	÷	●	●	+
280. Garcinia huillensis	●	●	●	●	●	+	●	●	●	●	+	●
281. Harungana madagascariensis	+	●	+	+	●	+	+	÷	÷	+	+	●
282. Allanblackia kimbiliensis	●	●	●	●	●	+	●	●	●	●	●	●
283. Mammea africana	●	●	+	●	●	●	●	●	●	●	●	●
284. Hypericum keniense	●	●	●	●	●	●	●	●	●	●	+	+
285. H.bequaertii	●	●	●	●	●	●	●	●	●	●	●	+

286. *H.revolutum*	●	●	●	●	●	●	●	●	●	●	+	+
287. *H.roeperianum*	●	●	●	●	●	●	●	●	●	●	+	●
288. *H.quartinianum*	●	●	●	●	●	●	●	●	●	●	+	●
289. *Anthocleista zambesiaca*	●	●	●	+	●	+	●	●	●	●	●	+
290. *A.vogelii*	●	●	●	+	●	+	+	●	●	●	●	●
291. *A.schweinfurthii*	+	+	●	●	●	+	●	÷	+	●	●	●
292. *Strychnis mitis*	+	÷	+	+	+	●	+	÷	÷	+	●	●
293. *Afrocrania volkensii*	●	●	●	●	●	●	●	●	●	●	+	+
294. *Memecylon jasminoides*	●	÷	+	+	÷	+	÷	●	÷	÷	●	●
295. *Memecylon sp.*	●	●	●	●	●	+	●	●	●	●	●	●
296. *Dichaetanthera corymbosa*	●	●	●	+	●	+	÷	●	●	●	●	●
297. *Mallotus oppositifolius*	+	●	+	●	●	●	●	●	+	÷	●	●
298. *Cassipourea malosana*	●	●	●	●	●	●	●	●	●	●	+	●
299. *C.ruwensorensis*	+	÷	+	+	●	+	+	÷	+	+	●	●
300. *C.congoensis*	●	●	●	+	●	+	÷	●	●	●	●	●
301. *C.gummiflua*	+	●	+	+	+	+	÷	÷	+	+	●	+
302. *Lasiodiscus mildbraedii*	+	+	+	+	+	●	●	●	+	+	●	●
303. *Mitragyna stipulosa*	●	÷	+	●	+	●	÷	●	●	●	●	●
304. *M.rubrostipulata*	+	●	●	+	●	+	÷	÷	+	●	●	●
305. *Nauclea diderrichii*	●	+	●	●	●	●	●	●	●	●	●	●
306. *Oxyanthus unilocularis*	●	+	●	●	●	●	+	●	●	●	●	●
307. *Pauridiantha callicarpoides*	●	●	●	+	●	+	÷	+	●	●	●	●
308. *P.viridiflora*	●	●	●	●	●	●	●	●	+	●	●	●
309. *Coffea liberica*	●	+	●	+	●	●	●	+	●	●	●	●
310. *Craterispermum laurinum*	+	●	+	+	●	●	●	÷	÷	●	●	●
311. *Vangueria apiculata*	●	●	+	+	+	●	÷	●	●	÷	+	●
312. *Morinda lucida*	●	÷	+	●	+	●	●	●	+	●	●	●
313. *Bertiera racemosa*	●	●	●	●	●	●	●	●	+	●	●	●
314. *Coffea canephora*	+	●	+	●	●	●	÷	÷	+	÷	●	●
315. *Belonophora pypoglauca*	+	÷	+	+	+	●	●	●	+	+	●	●

316. Dictyandra arborescens	+	•	+	+	+	•	+	•	÷	+	•	•
317. Oxyanthus speciosus	+	•	+	+	+	+	÷	÷	÷	÷	•	+
318. Pavetta insignis	•	•	+	+	•	•	•	÷	•	+	•	•
319. Tarenna pavettoides	+	+	+	+	•	+	+	•	÷	•	•	•
320. Galiniera coffeoides	+	•	+	+	•	+	+	÷	•	•	+	+
321. Psychotria megistosticta	+	+	•	+	•	+	•	•	+	•	•	+
322. Rothmannia urcelliformis	+	÷	+	+	+	•	+	÷	•	+	+	•
323. R. whitfieldii	+	•	+	•	•	•	•	•	•	•	•	•
324. Heinsenia diervilleiodes	•	•	•	+	•	•	•	•	•	•	+	•
325. Canthium vulgare	+	•	+	+	•	+	+	÷	÷	•	•	•
326. Aidia micrantha	•	•	•	+	•	+	+	÷	÷	•	•	•
327. Coffea eugenioides	+	•	+	+	•	•	+	•	+	•	•	•
328. Xymalos monospora	+	•	•	+	•	+	+	÷	÷	•	+	÷
329. Cassine buchananii	+	•	•	÷	•	•	÷	+	•	•	+	•
330. Catha edulis	•	•	•	•	•	•	•	•	•	•	+	•
331. Buddleia polystachya	•	•	•	•	•	•	•	•	•	•	+	•
332. Schrebera arborea	+	÷	+	+	+	•	+	•	+	+	•	•
333. Olea welwitschii	+	•	+	+	+	+	+	÷	+	÷	+	•
334. O.hochstetteri	•	•	•	+	•	+	•	•	•	•	+	+
335. O.africana	•	•	•	•	•	+	•	•	+	•	+	+
336. Linociera johnsonii	+	+	+	+	+	+	+	÷	+	+	+	•
337. L.latipetala	+	•	+	+	•	•	•	+	•	•	+	•
338. Olinia usambarensis	•	•	•	•	•	+	•	•	•	•	+	•
339. Syzygium guineense	•	•	•	•	•	+	•	•	+	•	+	•
340. S.cordatum	•	•	•	•	•	+	•	•	+	•	+	•
341. Eugenia bukobensis	•	•	•	+	•	•	•	•	÷	•	•	•
342. Memecylon myrianthum	•	•	•	•	•	•	•	•	+	÷	•	•
343. Balenites wilsoniana	+	÷	+	+	+	•	+	÷	÷	+	•	•
344. Erthrina excelsa	+	+	+	+	+	•	+	•	÷	÷	•	•
345. Erythrina sp.	•	÷	+	•	•	•	•	•	•	•	•	•

346. *Balsamocitrus dawei*	+	•	+	+	+	•	+	•	•	+	•	•
347. *Teclea nobilis*	+	+	+	+	+	+	+	÷	+	+	+	÷
348. *T.grandifolia*	•	÷	+	•	+	•	•	•	•	•	•	•
350. *Diphasia angolensis*	•	•	•	•	•	•	•	•	•	+	•	•
351. *Allophylus abyssinicus*	+	•	•	•	•	+	•	•	•	•	+	+
352. *A.macrobotrys*	+	÷	+	•	÷	+	÷	÷	•	+	+	•
353. *A.dummeri*	+	+	+	+	•	•	÷	•	÷	+	•	•
354. *Eudenia eminens*	+	•	+	+	•	•	+	÷	•	÷	•	•
355. *Ritchiea albersii*	+	÷	+	+	+	+	÷	÷	+	+	+	+
356. *Cussonia holstii*	•	+	•	•	•	•	•	•	•	•	•	•
357. *C.spicata*	•	•	•	•	•	•	•	•	•	•	+	•
358. *Schefflera abyssinica*	•	•	•	•	•	•	•	•	•	•	+	•
359. *S.volkensii*	•	•	•	•	•	•	•	•	•	•	+	•
360. *S.barteri*	•	÷	•	+	•	+	÷	•	+	•	•	•
361. *S.polyscidia*	•	•	•	•	•	•	•	•	•	•	•	+
362. *Ricinodentron heudelotii*	•	÷	+	+	+	•	•	•	•	+	•	•
363. *Bombax buonopozense*	•	+	+	+	+	•	•	•	•	+	•	•
364. *Myrianthus arboreus*	+	•	+	+	+	+	+	•	•	+	•	•
365. *M.holstii*	+	•	•	+	•	+	÷	÷	•	•	•	÷
366. *Musanga cecropioides*	•	+	+	•	•	+	•	•	•	÷	•	•
367. *M.leo-errerae*	•	•	•	+	•	+	•	•	•	•	•	•
368. *Mitex amboniensis*	+	•	+	+	+	•	÷	•	÷	+	•	•
369. *Markhamia platycalyx*	+	÷	+	+	+	+	+	÷	+	+	•	÷
370. *Spathodea campanulata*	+	÷	+	+	+	•	+	÷	+	÷	+	÷
371. *Kigelia africana*	+	÷	+	+	+	•	+	÷	÷	÷	+	+
372. *Fagaropsis angolensis*	+	•	+	+	+	•	+	÷	•	+	+	+
373. *Fagara macrophyllla*	+	÷	•	+	+	+	+	÷	•	•	+	•
374. *F.rubescens*	•	•	+	+	+	+	÷	•	÷	+	•	•
375. *F.leprieurii*	+	÷	+	•	•	+	+	÷	+	+	•	•
376. *F.mildbraedii*	•	•	•	+	•	+	•	•	•	•	•	•

Species	1	2	3	4	5	6	7	8	9	10	11	12
377. Clausena anisata	+	÷	+	+	+	+	+	÷	+	÷	●	+
378. Citropsis articulata	●	●	●	●	÷	●	●	÷	÷	÷	●	●
379. Schrebera alata	●	●	●	●	●	●	●	●	●	●	+	●
380. Bersama abyssinica	+	+	+	+	+	+	+	÷	+	+	+	+
381. Hagenia abyssinica	●	●	●	●	●	+	●	●	●	●	+	+
382. Harrisonia abyssinica	+	+	●	+	+	●	+	●	●	+	●	÷
383. Hannoa longipes	+	●	●	+	●	+	●	+	●	●	●	●
384. Polyscias fulva	+	●	●	+	●	+	+	÷	●	÷	●	÷
385. Pseudospondias microcarpa	+	+	+	+	+	+	+	÷	+	+	●	●
386. Tricoscypha submontana	●	●	●	+	●	●	÷	●	●	●	●	●
387. Lannea welwitschii	●	÷	+	+	+	÷	●	●	●	+	●	●
388. Antrocaryon micraster	●	●	+	●	●	●	●	●	●	+	●	●
389. Canarium schweinfurthii	●	÷	+	+	●	+	●	●	+	÷	●	●
390. Ekerbergia senegalensis	+	●	+	+	+	●	●	●	÷	÷	●	●
391. E. capensis	●	●	+	●	●	+	●	÷	÷	●	+	●
392. Trichilia dregeana	+	+	+	+	+	●	+	÷	+	÷	+	●
393. T.martineaui	●	●	+	÷	+	+	÷	●	●	+	●	●
394. T.prieureana	+	●	+	+	+	●	●	÷	●	+	●	●
395. T.rubescens	+	●	+	+	●	●	+	+	+	÷	●	●
396. Lepidotrichilia volkensii	+	●	●	+	●	+	÷	÷	÷	●	+	+
397. Carapa grandiflora	+	●	●	+	●	+	+	÷	+	●	●	●
398. Leplaea mayombensis	●	●	●	●	●	+	●	●	●	●	●	●
399. Entandrophragma utile	●	●	+	●	+	●	●	●	●	+	●	+
400. E.cylindricum	+	÷	+	●	+	+	●	●	●	●	●	●
401. E.angolense	+	●	+	÷	+	●	÷	●	●	+	●	●
402. E.excelsum	+	●	●	+	●	+	+	+	÷	●	+	÷
403. Khaya anthotheca	●	+	+	●	+	●	●	●	●	●	●	●
404. K.grandifoliola	●	●	+	●	●	●	●	●	●	●	●	●
405. Guarea cedrata	●	+	+	●	+	●	÷	●	+	+	●	●
406. Lovoa trichilioides	●	●	+	●	+	●	+	●	+	+	●	●

407. L.swynnertonii	+	÷	•	+	+	+	+	÷	•	÷	•	•
408. Turraeanthus africanus	+	÷	•	•	•	•	÷	+	•	•	•	•
409. Majidea fosteri	+	•	+	+	+	•	+	•	+	+	•	•
410. Deinbollia fulvo-tomentella	•	•	•	•	•	•	+	•	•	•	•	•
411. D.kilimandscharica	•	•	+	+	•	•	•	•	•	•	•	•
412. Lychnodiscus cerospermus	+	÷	+	+	+	•	•	•	+	+	•	•
413. Lecaniodiscus fraxinifolius	•	•	•	•	•	•	•	•	÷	+	•	•
414. L.cupnioides	•	+	•	•	•	•	÷	•	•	•	•	•
415. Blighia welwitschii	+	÷	+	•	•	•	•	÷	+	÷	•	•
416. Zanha golungensis	+	•	+	•	+	+	÷	•	•	÷	•	•
417. Pancovia sp.near P.turbinata	+	•	•	+	+	+	÷	÷	÷	÷	•	•
418. Melanodiscus sp.	•	•	+	•	+	•	•	÷	•	+	•	•
419. Blighia unijugata	+	÷	+	+	+	÷	+	÷	÷	+	•	•
420. Aphania senegalensis	+	÷	+	+	+	+	÷	÷	•	÷	•	•
421. Connarus longistipitatus	•	÷	+	+	•	•	•	÷	+	•	•	•
422. Cnestis ugandensis	+	÷	+	•	•	•	÷	•	•	+	•	•
423. Mildbraediodendron excelsum	•	÷	+	+	+	•	•	•	•	+	•	•
424. Cassia mannii	•	+	+	+	+	•	•	•	•	•	•	•
425. Dialium excelsum	•	+	+	•	+	•	•	+	•	•	•	•
426. Afzelia bipindensis	•	+	•	•	•	•	•	•	•	•	•	•
427. Baikiaea insignis	•	•	•	+	+	•	÷	•	+	+	•	•
428. Cynometra alexandri	+	+	+	+	+	+	+	•	•	•	•	•
429. Craibia brownii	+	÷	+	+	•	•	•	÷	•	+	+	•
430. Millettia dura	+	÷	•	+	•	+	+	÷	•	•	•	•
431. M.eetveldeana	•	+	•	•	•	•	•	•	•	•	•	•
432. M.psilopetala	•	+	•	•	•	+	•	•	•	•	•	•
433. Erythrophleum suaveolens	•	•	+	•	+	•	•	•	•	+	•	•
434. Piptadeniastrum africanum	+	•	+	+	+	+	+	•	•	÷	•	•
435. Newtonia buchananii	+	•	•	+	+	+	+	÷	•	•	•	•
436. Cathormion altissimum	•	•	+	•	+	•	•	•	•	•	•	•

Species	1	2	3	4	5	6	7	8	9	10	11	12
437. *Acacia kirkii*	●	●	●	●	●	●	●	●	÷	●	●	●
438. *Dichrostachys glomerata*	●	+	+	●	+	●	●	●	●	+	●	●
439. *Tetrapleura tetraptera*	+	+	+	●	+	●	+	÷	+	+	●	●
440. *Parkia filicoidea*	+	÷	+	+	+	+	+	●	+	÷	●	●
441. *Albizia ferruginea*	+	÷	+	+	+	●	+	÷	●	+	●	●
442. *A.coriatia*	●	÷	+	+	+	+	+	÷	●	÷	●	●
443. *A.glaberrima*	+	÷	+	●	+	●	●	●	+	+	●	●
444. *A.gummifera*	+	●	●	+	●	+	+	●	+	+	+	●
445. *A.adianthifolia*	●	●	●	+	●	+	+	÷	●	●	●	÷
446. *A.grandibracteata*	+	÷	+	+	+	+	+	÷	●	+	+	●
447. *A.zygia*	+	÷	+	+	+	●	+	+	+	+	●	●

Key:

+   denotes species presence recorded in the literature and/or at Makerere University herbarium prior to this study;

÷ denotes species recorded for the first time during this study;

●   denotes species not recorded.

Forests are designated as:

1 = Kibale
2 = Semliki
3 = Budongo
4 = Kalinzu-Maramagambo
5 = Bugoma
6 = Bwindi,
7 = Kasyoha-Kitomi
8 = Itwara
9 = Sango Bay
10 = Mabira
11 = Mount Elgon
12 = Rwenzori

Tree species numbers those used in Hamilton (1981). Non-forest species are excluded (Howard 1991).

Exotic Trees of Uganda

Species Name	Family	Origin
Acacia baileyan	Mimosaceae	
Acacia decurrens	Mimosaceae	
Acacia mearnsii	Mimosaceae	Australia
Acacia elata	Mimosaceae	Australia
Acacia farnesiana	Mimosaceae	Tropical America
Acacia melanoxylon	Mimosaceae	Australia
Acacia podalyriifolia	Mimosaceae	Australia
Acacia pruinosa	Mimosaceae	
Acacia pycnantha	Mimosaceae	
Acacia saligna	Mimosaceae	Australia
Achras zapotilla	Sapotaceae	Central America
Acrocarpus fraxinifolius	Caesalpiniaceae	India
Adansonia digitata	Bombacaceae	Sudan/drier parts of E. Africa
Adenanthera pavonina	Mimosaceae	Tropical Asia
Aegle marmelos	Rutaceae	
Afrormosia elata	Papilionaceae	Congo
Afzelia quanzensis	Caesalpiniaceae	
Albizzia chinensis	Mimosaceae	Tropical Asia
Albizzia falcata	Mimosaceae	Malaysia
Albizzia lebbeck	Mimosaceae	Tropical Asia and Africa
Albizzia odorantissima	Mimosaceae	Tropical Asia
Albizzia procera	Mimosaceae	India
Aleurites fordii	Euphorbiaceae	China
Aleurites molluccana	Euphorbiaceae	Malaysia & Pacific Islands

Species Name	Family	Origin
Amomis caryophyllata	Myrtaceae	Tropical America
Arnacardium occidentale	Anarcardicae	Tropical America
Annona cherimolia	Annonaceae	South America
Annona muricata	Annonaceae	India
Annona reticulata	Annonaceae	India
Annona squamosa	Annonaceae	India
Araucaria angustifolia	Araucariaceae	Brazil & Argentina
Araucaria araucana	Araucariaceae	Chile & Argentina
Araucaria bidwillii	Araucariaceae	Australia
Araucaria columnaris	Araucariaceae	New Caledonia &Polynesia
Araucaria cunninghamii	Araucariaceae	Australia & New Guinea
Araucaria excelsa	Araucariaceae	Norfolk Island
Archontophoenix alexandrae	Palmaceae	Australia
Areca catechu	Palmaceae	Tropical Asia
Areca concinna	Palmaceae	Ceylon
Arenga pinnata	Palmaceae	Malaysia
Artocarpus altilis	Moraceae	Polynesia
Artocarpus heterophyllus	Moraceae	Tropical Asia
Artocarpus integer	Moraceae	Malaysia
Averrhoa carambola	Oxalidaceae	Moluccas
Azadirachta indica	Meliaceae	India
Bauhinia purpurea	Caesalpiniaceae	India
Bauhinia racemosa	Caesalpiniaceae	India
Bauhinia variegata	Caesalpiniaceae	India
Bischoffia javanica	Euphorbiaceae	Tropical Asia
Bixa orellana Bixaceae	Tropical	America
Bolusanthus speciosus	Papilionaceae	South Africa

Species Name	Family	Origin
Bombax malabaricum	Bombacaceae	India
Brachychiton acerifolium	Sterculiaceae	Australia
Brachychiton populneum	Sterculiaceae	Australia
Broussonetia papyrifera	Moraceae	Pacific Islands
Brownea ariza	Caesalpiniaceae	Tropical America
Bursera delpechiana	Burseraceae	Mexico
Caesalpinia coriaria	Caesalpiniaceae	Tropical America
Caesalpinia paucijuga	Caesalpiniaceae	India
Caesalpinia sappan	Caesalpiniaceae	India & Burma
Caesalpinia spinosa	Caesalpiniaceae	Bolivia & Venezuela
Callistemon	Myrtaceae	Australia
Callitris calcarata	Cupressaceae	Australia
Callitris cupressiformis	Cupressaceae	Australia
Callitris robusta	Cupressaceae	Australia
Calophyllum inophyllum	Guttiferae	S.E Asia
Cananga odorata	Annonaceae	Malaya
Carica candamarcensis	Caricaceae	Columbia & Ecuador
Carica papaya	Caricaceae	
Caryota mitis	Caricaceae	S.E Asia
Caryota urens	Caricaceae	Tropical Asia
Casimiroa edulis	Rutaceae	Central America
Cassia fistula	Caesalpiniaceae	India
Cassia fruticosa	Caesalpiniaceae	Tropical America
Cassia grandis	Caesalpiniaceae	Tropical America
Cassia javanica	Caesalpiniaceae	Malaysia
Cassia leiandra	Caesalpiniaceae	Tropical America
Cassia marginata	Caesalpiniaceae	India
Cassia multijuga	Caesalpiniaceae	Tropical America

Species Name	Family	Origin
Cassia nodosa	Caesalpiniaceae	Bengal & Malaysia
Cassia renigera	Caesalpiniaceae	Burma
Cassia siamea	Caesalpiniaceae	S. Asia
Cassia spectabilis	Caesalpiniaceae	Tropical America
Castanospermum australe	Papilionaceae	Australia
Castilla elastica	Moraceae	Mexico
Casuarina cunninghamiana	Casuarinaceae	Australia
Casuarina equisetifolia	Casuarinaceae	E. African Coast
Casuarina suberosa	Casuarinaceae	
Cecropia peltata	Moraceae	Tropical America
Cedrela odorata	Meliaceae	Central America
Cieba pentandra	Bombacaceae	Tropical America
Ceratonia siliqua	Caesalpiniaceae	
Chamaecyparis obtusa	Cupressaceae	Japan
Chamaerops humilis	Palmaceae	Europe
Chorisia speciosa	Bombacaceae	Brazil
Chrysalidocarpus lutescens	Palmaceae	Madagascar
Chrysophyllum cainito	Sapotaceae	Central America
Chinchona	Rubiaceae	S. America
Cinnamomum camphora	Lauraceae	Eastern Asia
Cinnamomum zeylanicum	Lauraceae	Tropical Asia
Citrus aurantiifolia	Rutaceae	Malaysia
Citrus aurantium	Rutaceae	China
Citrus limonia	Rutaceae	Asia
Citrus maxima	Rutaceae	Malaysia
Citrus medica	Rutaceae	India & China
Citrus reticulata	Rutaceae	China
Citrus simensis	Rutaceae	China

Species Name	Family	Origin
Coccoloba uvifera	Polygonaceae	South America
Cocos nucifera	Palmaceae	Tropical America
Cocos weddelliana	Palmaceae	Brazil
Coffea arabica	Rubiaceae	Ethiopia
Coffea congensis	Rubiaceae	Tropical Africa
Coffea liberica	Rubiaceae	West Africa
Coffea stenophylla	Rubiaceae	West Africa
Cola acuminata	Sterculiaceae	West Africa
Cordia myxa	Boraginaceae	India
Couroupita guianensis	Lecythidaceae	Surinam & Trinidad
Craibia elliotii	Papilionaceae	Kenya & Tanzania
Crescentia cujete	Bignoniaceae	Tropical America
Croton tiglium	Euphorbiaceae	Tropical Asia
Cryptomeria japonica	Taxodiaceea	Japan
Cupressus greene	Cupressaceae	Arizona
Cupressus funebris	Cupressaceae	China
Cupressus lusitanica	Cupressaceae	Mexico
Cupressus macrocarpa	Cupressaceae	California
Cupressus sempervirens	Cupressaceae	Mediterranean
Cupressus torulosa	Cupressaceae	Bhutan & Himalayas
Cycas circinalis	Cycadaceae	Tropical Asia & Africa
Cycas revoluta	Cycadaceae	Japan
Cydonia oblonga	Rosaceae	
Cyrtostachys lakka	Palmaceae	Malaysia
Dalbergia latifolia	Papilionaceae	India
Dalbergia sissoo	Papilionaceae	India
Delonix regia	Caesalpiniaceae	Madagascar
Derris microphylla	Papilionaceae	Burma & India

Species Name	Family	Origin
Didymopanax morototoni	Araliaceae	S. America & Trinidad
Dillenia indica	Dilleniceae	Tropical Asia
Diospyros discolour	Ebenaceae	Philippines
Dovyalis caffra	Flacourtiaceae	S. Africa
Dovyalis hebecarpa	Flacourtiaceae	Ceylon
Durio zibethinus	Bombacaceae	Malaysia
Dyera costulata	Apocynaceae	Malaysia
Elaeocarpus grandis	Tiliaceae	Australia
Emblica officinalis	Euphorbiaceae	Tropical Asia
Eriobotrya japonica	Rosaceae	China & Japan
Erythrina crista-galli	Papilionaceae	Brazil
Erythrina glauca	Papilionaceae	Venezuela
Erythrina umbrosa	Papilionaceae	S. America
Erythrina variegata	Papilionaceae	Tropical Asia
Erythrina velutina	Papilionaceae	Venezuela
Eucalyptus bicolor	Myrtaceae	Australia
Eucalyptus bosistoana	Myrtaceae	Australia
Eucalyptus botryoides	Myrtaceae	Australia
Eucalyptus camaldulensis	Myrtaceae	Australia
Eucalyptus citriodora	Myrtaceae	Australia
Eucalyptus cloeziana	Myrtaceae	Australia
Eucalyptus crebra	Myrtaceae	Australia
Eucalyptus deglupta	Myrtaceae	Australia
Eucalyptus ficifolia	Myrtaceae	Australia
Eucalyptus globulus	Myrtaceae	Australia
Eucalyptus grandis	Myrtaceae	Australia
Eucalyptus albens	Myrtaceae	Australia
Eucalyptus leucoxylon	Myrtaceae	Australia

Species Name	Family	Origin
Eucalyptus maculata	Myrtaceae	Australia
Eucalyptus maideni	Myrtaceae	Australia
Eucalyptus melliodora	Myrtaceae	Australia
Eucalyptus microcorys	Myrtaceae	Australia
Eucalyptus paniculata	Myrtaceae	Australia
Eucalyptus planchoniana	Myrtaceae	Australia
Eucalyptus punctata	Myrtaceae	Australia
Eucalyptus resinifera	Myrtaceae	Australia
Eucalyptus robusta	Myrtaceae	Australia
Eucalyptus rudis	Myrtaceae	Australia
Eucalyptus saligna	Myrtaceae	Australia
Eucalyptus salmonophloia	Myrtaceae	Australia
Eucalyptus scabra	Myrtaceae	Australia
Eucalyptus siderophloia	Myrtaceae	Australia
Eucalyptus sideroxylon	Myrtaceae	Australia
Eucalyptus tereticornis	Myrtaceae	Australia
Eugenia floribunda	Myrtaceae	India
Eugenia myrtifolia	Myrtaceae	Australia
Eugenia uniflora	Myrtaceae	S. America
Feijoa sellowiana	Myrtaceae	Brazil
Feronia limonica	Rutaceae	India
Ficus benghalensis	Moraceae	India
Ficus benjamina	Moraceae	Malaysia
Ficus religiosa	Moraceae	India
Filicium decipiens	Sapindaceae	Kenya, Tanzania & Asia
Firmiana plantanifolia	Sterculiaceae	China & Japan
Flacourtia jangomas	Flacourtiaceae	Tropical Asia
Fraxinus berlandieriana	Oleaceae	Mexico

Species Name	Family	Origin
Fraxinus viridis	Oleaceae	Americas
Garcinia mangostana	Guttiferae	Malaysia
Gleditsia triacanthos	Caesalpiniaceae	N. America
Gliricidia sepium	Papilionaceae	Tropical America
Gmelina arborea	Verbenaceae	Tropical Asia
Grevillea banksii	Proteaceae	Australia
Grevillea robusta	Proteaceae	Australia
Haematoxylon campechianum	Caesalpiniaceae	Central America
Hevea brasiliensis	Euphorbiaceae	Amazon
Hovenia dulcis	Rhamnaceae	China
Hura crepitans	Euphorbiaceae	Tropical America
Hydnocarpus anthelmintica	Flacourtiaceae	Siamese
Hydnocarpus laurifolia	Flacourtiaceae	India
Hyemenaea courbaril	Caesalpiniaceae	India
Ixora parviflora	Rubiaceae	India
Jacaranda mimosifolia	Bignoniaceae	Brazil
Jambosa caryophyllus	Myrtaceae	Molucca Islands
Jambosa jambos	Myrtaceae	Tropical Asia
Juniperus bermudiana	Cupressaceae	Bermudans
Kleinhovia hospita	Sterculiaceae	Malaya
Kopsia fruticosa	Apocynaceae	Malaysia
Largerstroemia indica	Lythraceae	China
Largerstroemia speciosa	Lythraceae	Tropical Asia
Lagunaria patersonii	Malvaceae	Australia
Lansium domesticum	Meliaceae	Malaysia
Leptospermum scoparium	Myrtaceae	Australia
Leucaena glauca	Mimosaceae	Tropical America
Leucaena leucocephala	Mimosaceae	Tropical America

Species Name	Family	Origin
Leucaena pulverulenta	Mimosaceae	Mexico
Leucaena glabrata	Mimosaceae	Mexico
Licuala grandis	Palmaceae	New Britain
Litchi chinensis	Sapindaceae	China
Lonchocarpus sepium	Papilionaceae	Central America
Lonchocarpus sericeus	Papilionaceae	Tropical America
Lysidice rhodostegia	Caesalpiniaceae	China
Macadamia ternifolia	Proteaceae	Australia
Magnolia grandiflora	Magnoliaceae	N. America
Malpighia glabra	Malpighiaceae	Tropical America
Malus pumila	Rosaceae	
Mangifera indica	Anacardiaceae	Tropical Asia
Manihot glaziovii	Euphorbiaceae	Brazil
Melaleuca leucadendron	Myrtaceae	Australia
Melia azedarach	Meliaceae	India
Melia dubia	Meliaceae	Asia & Australia
Melicocca bijuga	Sapindaceae	S. America
Michelia champaca	Magnoliaceae	Tropical Asia
Michelia fuscata	Magnoliaceae	China
Millingtonia hortensis	Bignoniaceae	Burma
Mimosa bracaatinga	Mimosaceae	Brazil
Moringa oleifera	Moringaceae	India & Arabia
Morus alba	Moraceae	
Myristica fragrans	Myriticaceae	Moluccas
Myroxylon balsamum	Papilionaceae	S. America
Nephelium lappaceum	Sapindaceae	Asia
Roedoxa	Palmaceae	Tropical America
Bombax kimuenzae	Bombacaceae	India

Species Name	Family	Origin
Palaquium javense	Sapotaceae	Java
Pandanus utilis	Pandanaceae	Madagascar
Pandanus veitchii	Pandanaceae	Polynesia
Parkia javanica	Mimosaceae	Tropical Asia
Parkinsonia aculeata	Caesalpiniaceae	Tropical America
Parmentiera cerifera	Bignoniaceae	Panama
Peltophorum dasyrhachis	Caesalpiniaceae	E. Indies
Peltophorum inerme	Caesalpiniaceae	Tropical Asia
Pentandesma butyracea	Guttiferae	West Africa
Persea americana	Lauraceae	Tropical America
Phoenix dactylifera	Palmaceae	N. Africa & Arabia
Phyllunthus acidus	Euphorbiaceae	Madagascar
Phytelephas macrocarpa	Palmaceae	Columbia
Phytolacca dioica	Phytolaccaceae	S. America
Pimenta officinalis	Myrtaceae	Tropical America
Pinus canariensis	Pinaceae	Canary Is.
Pinus caribaea	Pinaceae	USA, West Indies & Honduras
Pinus echinata	Pinaceae	USA
Pinus halepensis	Pinaceae	Mediterranean
Pinus khasya	Pinaceae	Burma
Pinus leiophylla	Pinaceae	USA & Mexico
Pinus merkusii	Pinaceae	S. Asia
Pinus montezumae	Pinaceae	Mexico
Pinus oocarpa	Pinaceae	Central America
Pinus patula	Pinaceae	Mexico
Pinus pinaster	Pinaceae	Mediterranean
Pinus pseudostrobus	Pinaceae	Mexico
Pinus radiata	Pinaceae	California

Species Name	Family	Origin
Pinus roxburghii	Pinaceae	Himalayas
Pinus strobus	Pinaceae	USA & Canada
Pinus taeda	Pinaceae	USA
Pithecellobium dulce	Mimosaceae	Tropical America
Polyalthia longfolia	Annonaceae	Ceylon
Pongamia pinnata	Papilionaceae	India
Populus deltoides	Salicaceae	N. America
Prosopsis chilensis	Mimosaceae	India & C. America
Prunus cerasoides	Rosaceae	
Prunus domestica	Rosaceae	
Prunus laurocerasus	Rosaceae	
Prunus persica	Rosaceae	
Prunus serotina	Rosaceae	N. America
Psidium cattleianum	Myrtaceae	Brazil
Psidium guajava	Myrtaceae	Tropical America
Ptychosperma macarthuri	Palmaceae	Australia
Punica granatum	Punicaceae	India & Persia
Putranjiva roxburghii	Euphorbiaceae	India
Pyrus communis	Rosaceae	
Pyrus malus	Rosaceae	
Quercus robur	Fagaceae	Mediterranean
Quercus suber	Fagaceae	Mediterranean
Ravenala madagascariensis	Musaceae	Madagascar
Roystonea oleracea	Palmaceae	Tropical America
Roystonea regia	Palmaceae	Tropical America
Salix babylonica	Salicaceae	Caucasian Region
Samanea saman	Mimosaceae	C. & S. America
Santalum album	Santalaceae	India

Species Name	Family	Origin
Sapindus saponaria	Sapindaceae	Tropical & sub-tropical America
Sapindus trifoliatus	Sapindaceae	India
Sapium sebiferum	Euphorbiaceae	China
Saraca indica	Caesalpiniaceae	India
Schinus mole	Anacardiaceae	The Andes
Schinus terebinthifolius	Anacardiaceae	Brazil
Schizolobium excelsum	Caesalpiniaceae	Brazil
Schotia latifolia	Caesalpiniaceae	S. Africa
Solanum macranthum	Solanaceae	Brazil
Sterculia foetida	Sterculiaceae	Tropical Asia
Styrax benzoin	Styracea	Malaysia
Swietenia macrophylla	Melliaceae	Central America
Swietenia mahogoni	Melliaceae	Florida & W. Indies
Syncarpia glomulifera	Myrtaceae	Australia
Syzygium cumini	Myrtaceae	Asia & E. A Coast
Tabebuia pentaphylla	Bignoniaceae	Tropical America
Tabebuia rosea	Bignoniaceae	Tropical America
Tercoma stans	Bignoniaceae	Tropical America
Tectona grandis	Verbenaceae	Tropical Asia
Terminalia belerica	Combretaceae	India & Malaysia
Terminalia catappa	Combretaceae	India
Thea sinensis	Theaceae	India
Theobroma cacoa	Sterculiaceae	S.& C America
Thespesia populnea	Malvaceae	Tropical Coasts
Thuja occidentalis	Cupressaceae	N. America
Tipuana tipu	Papilionaceae	Brazil
Toona ciliata	Melliaceae	Tropical Asia
Toona serrata	Melliaceae	Himalayas

Species Name	Family	Origin
Tristania conferta	Myrtaceae	Australia
Tristania suaveolens	Myrtaceae	Australia
Vitex keniensis	Verbenaceae	Kenya
Widdringtonia juniperoides	Cupressaceae	S. Africa
Widdringtonia whytei	Cupressaceae	Malawi
Xylia xylocarpa	Mimosaceae	Tropical Asia

Source: Compiled from Research Plot Records 1998, FORI, Kampala

2
AGROFORESTRY

N. Wajja-Musukwe and M.I. Mbalule

In Uganda there are many traditional land use practices that involve production of trees and agricultural crops on the same piece of land. These land use systems are called agroforestry which is a new name for old practices. In most cases, the major objective of these practices is not tree production but food production. The trees are deliberately retained for various cultural and economic reasons.

Definition of agroforestry

Agroforestry is a collective name for land-use systems and technologies in which woody perennials (trees, shrubs, palms, bamboos, etc.) are deliberately used on the same land-management units as agricultural crops and/or animals in some form of spatial arrangement or temporal sequence (Lundgren and Raintree, 1992). Therefore agroforestry is an approach to land use involving a deliberate and purposeful combination of trees with crops and/or animals.

Importance of agroforestry

Increasing pressure on land, resulting from a rapidly growing population, has caused problems of deforestation and environmental degradation. For example, unregulated exploitation and large scale encroachment has robbed the Mabira Forest Reserve in Mukono district of its most valuable species, reducing them to only 0.06% of the growing stock. Consequently, it has become vital to identify appropriate land-use approaches that lead to production of multipurpose outputs and ensure the sustainability of the production base.

Agroforestry, as a land use approach, is important because, by integrating multipurpose trees with crop production, the problems of poor agricultural production, worsening wood shortages and environmental degradation can be addressed. Furthermore, agroforestry practices are seen as an opportunity to take pressure off the remaining natural forests and to increase the diversity of vegetation on existing farms.

The basic and overriding characteristic of agroforestry tree species is that they are multipurpose, providing more than one product or service. A tree can be multipurpose in two ways:

(i)   By yielding more than one product at the same time, e.g.,

    (a)   *Calliandra calothyrsus* and *Gliricidia sepium* if planted as living fences can provide firewood, fodder, and green manure for agricultural crops all at the same time. In Kabale and Kisoro districts, hedgerows of *Calliandra* planted for control of run-off also provide stakes for climbing beans.

(b)   The jackfruit (*Artocarpus heterophyllus)* provides fruit which is a source of food and income in several areas of Uganda while in very dry periods its leaves can be used as fodder.

(c)   *Ficus natalensis,* whether interplanted in perennial crops or along farm boundaries, provides firewood and fodder at the same time. This is a common practice in the banana-coffee lakeshore land use system.

(ii)   By yielding different products depending on how it is managed, e.g., *Calliandra* can be planted in blocks to provide fuelwood or stakes; alternatively, it can be cut back frequently to provide fodder or green manure.

On farms, trees provide a diversity of products and service functions which justify their inclusion in crop land.

Products from trees include
(i) Raw materials for industry

(a) wood for furniture, carving, hand tools, musical instruments, weapons, timber for building. Rattan cane is harvested from Mabira, Budongo and other forest reserves for easy chairs and stools.

(b) fibres, ropes, baskets and clothing; Leaves of *Phoenix reclinata,* a wild date palm locally called 'Mukindu' are used for floor mats, hats and other products, while bark cloth is made from the bark of *Ficus natalensis.*

(c) industrial chemicals (gums, dyes, medicines, essential oils, rubber, etc).

(ii) Food for people (fruits, nuts, oils, honey, spices, mushrooms, etc.).

(iii) Feed for livestock (leaves, fruits, pods, barks etc.). The largest proportion of fresh passion fruits in Kampala and other urban areas are collected from forest reserves.

(iv) Energy (firewood, charcoal, saw dust/chips, etc).

Services from trees include
(i) Live fencing (*Dovyalis caffra* is commonly used as a live fence around homesteads).

(ii) Shade for people, crops and animals. *Cordia africana* in Mbale and Bundibugyo districts, *Ficus natalensis* and *Albizia coriaria* in the banana-coffee lake shore land use system provide agricultural shade.

(iii) Shelter belts - including control of wind erosion and crop desiccation and protection of agricultural crops against damage by strong winds.

iv) Soil fertility replenishment through biological nitrogen fixation, recycling of nutrients within the soil, and control of water and soil run off. In Kabale district, N-fixing trees are planted on terraces to stabilise them and they are periodically cut and the prunings applied to the cropped terrace bench as green manure.

Classification of agroforestry systems and practices

Although the words systems and practices are widely used synonymously in agroforestry literature, a distinction can be drawn between them.

An agroforestry practice refers to the distinctive arrangement of components in space and time. On the other hand, an agroforestry system is a practice that is specific to a locality or region. Hence, a system embraces not only the plant species and their arrangement, management and socioeconomic functioning but also the environment, where they exist. Admittedly, the distinction between systems and practices is thin and vague and consequently, the terms are used synonymously.

Various types of agroforestry systems exist in different ecological regions of Uganda. Undoubtedly, the classification of agroforestry systems is a complex process.

Another term that is frequently used is agroforestry technology. It refers to an innovation or improvement, usually through scientific intervention, to either modify an existing system or practice, or develop a new one (Nair, 1993).

The most commonly used criteria for classifying agroforestry systems and practices are:

- Structure of the system (including the nature and spatial and temporal arrangement of components).

- Function of the system (output and role of components, particularly the woody component, e.g., soil conservation, wind break).

- Agroecological zones where the system exists (e.g., tropical highlands, arid and semi arid lands, etc.).

- Socioeconomic scales and levels of management of the system (whether low input, high input, commercial, intermediate or subsistence).

Classification based on structure of the system

The structural classification of agroforestry systems is based both on the nature of components involved and the arrangement thereof.

Based on the nature of components. In all agroforestry systems, there are three basic components that are managed by the land user; namely, woody perennials, agricultural crops including pastures and the animal. Based on the nature of combination of these components, there are three major categories:

- agrosilvocultural, in which agricultural crops (including shrubs, vines and tree crops) are intercropped with trees.

- Silvopastoral, in which trees are integrated with pastures and/or animals and

- agrosilvopastoral, where crops, pasture and/or animals are integrated.

In addition to these three broad categories, there are a few other systems such as woodlots which interact with other land use production components, apiculture with trees (apisilviculture/ entomoforestry) and integration of trees and shrubs with fish production (aquasilviculture).

Based on the arrangement of components. The arrangement of components refers to the plant components and their arrangement in multispecies combination in space and time. Spatial arrangement of plants in agroforestry mixtures may be in the form of dense mixed stands (as in home gardens), sparsely mixed stands (as in many silvopastoral systems) or zones (strips) of varying width (as in alley cropping and boundary planting). Temporal arrangements of plants in agroforestry may take the form of conventional shifting cultivation or some silvopastoral systems that involve rotation of grass leys with woody species.

Classification based on function of systems. In addition to having a productive function (yielding one or more products) all agroforestry systems have a service role (protecting and maintaining the production base). The latter attribute ensures sustainability, which differentiates agroforestry from other approaches to land use. On the basis of the degree of importance of a particular role, an agroforestry system can be classified as productive or protective. For example, tree planting on erosion-control structures contributes to control of runoff and erosion by stabilising earth structures through their root systems. Such systems are protective.

Ecological classification. Many agroforestry systems are characterised on the basis of ecological conditions where they occur. Thus, there are agroforestry systems, for example, the *Ficus* and coffee-banana system in central Uganda and Okoro county of Nebbi district. In general, there can be agrosilvoculture, silvopastoral or agrosilvopastoral systems in any of the ecological regions. Therefore, agroecological conditions alone can not be taken as a satisfactory basis for classification of agroforestry systems, but agroecological characteristics can be used as a basis for designing agroforestry systems (Nair, 1993).

Classification based on socioeconomic criteria. Another basis for classifying agroforestry systems are socioeconomic criteria such as scale of production and level of technology input and management. For example, agroforestry systems have been grouped into commercial, intermediate and subsistence systems by Lundgren (1982).

Commercial agroforestry systems are those in which production is the major aim of the system, on medium to large scale of production, with hired labour, and on private, corporate or government land. Examples include commercial production of shade-tolerant agricultural plantation crops such as coffee, cocoa and tea under shade trees as well as commercial grazing and ranching under large-scale timber and pulp plantations.

Subsistence agroforestry systems are those geared toward satisfying basic needs and the land is managed by family labour. Most of the agroforestry systems in different parts of developing countries fall in this category. Examples include the resource-depleting traditional shifting cultivation and the ecologically sound home gardens (Torquebiau, 1992).

Intermediate agroforestry systems lie between commercial and subsistence scales of production and management. Cash crops are grown to satisfy cash needs while food crops meet the family's subsistence needs. Most agroforestry systems based on plantation crops such as coffee, coconut and cocoa fall in this category.

Like classification based on ecological conditions, categorizing agroforestry systems according to socioeconomic criteria can not be used as the primary basis for classifying the systems.

In summary, while each of these criteria has its own merits and limitations and, therefore, no single criterion can be chosen as universally applicable, Nair (1993) has suggested that the complexity of agroforestry classification can be reduced if the structural and functional aspects of the system are taken as the criteria for categorizing the systems and agroecological and socioeconomic aspects as the basis for further grouping or stratification.

Distribution of agroforestry systems

Although not documented, numerous traditional agroforestry systems/practices have existed in Uganda for a long time. In an attempt to develop an appropriate agroforestry research agenda for Uganda, Djimde and Hoekstra (1988) identified nine land use systems on the basis of which agroforestry potentials and agroforestry research needs for different land use systems were suggested.

Generally, the ecological conditions of an area is the major factor that determines the distribution and extent of adoption of specific agroforestry systems. Thus, areas with similar ecological conditions tend to have structurally similar agroforestry systems.

The distribution of different agroforestry systems in Uganda is presented in this section.

Agrosilvocultural systems

(i) Shifting cultivation
Shifting cultivation refers to farming in which land under natural vegetation is cleared, cropped with agricultural crops for a few years (2-3 years) and then left for a much longer fallow period (10-20 years) for a natural vegetation to regenerate.

The practice is common in forest areas where a patch of forest is cleared during the dry period, burning the debris in situ and planting crops at the onset of rains. In Uganda, communities that live near natural forests, e.g., Budongo and Mabira forest reserves practise shifting cultivation. The system is stable and ecologically sound, but owing to increasing population pressure, the fallow periods have reduced, resulting in serious soil erosion and a decline in the soil's fertility and productivity.

(ii) Taungya

The Taungya system consists of growing annual agricultural crops along with forestry tree crops during the early years of establishment of a forestry plantation. The land belongs to the Forestry Department, which allows the subsistence farmers to grow agricultural crops. On their part, the farmers are required to tend the tree crops seedlings and retain all the agricultural produce. As the tree crops grow and expand their canopy, they shade the crops. This, coupled with declining soil fertility and increased weed infestation, make crop production non productive and the farmers cease intercropping.

The Taungya system is a popular and successful agroforestry approach to establishing forest plantations in Uganda. In the peri-urban plantations of Namanve, Kyewaga, and Kitubulu in Mpigi district, the Forest Department employs taungya system to establish *Eucalyptus* plantations. *Cupressus lusitanica* plantations are established in a similar manner in Mafuga Forest Reserve.

(iii) Home gardens

The term home garden has been used loosely to mean several practices, such as growing vegetables behind houses. In agroforestry terminology, however, home gardens are mixed plantings where multipurpose trees and shrubs are grown in intimate association with annual and perennial crops and livestock within the compound of individual houses and under the management of family labour (Fernandes and Nair, 1986).

Home gardens are multistoreyed in structure, meaning that there are several layers (canopies) of plants growing to different heights in the system. The lowest level often consists of vegetables or root crops (e.g. yams); the second level includes fast growing trees or crops such as bananas, coffee, cocoa; a third, higher level may consist of large trees that provide fruit, timber, fuelwood and shade.

In Uganda, home gardens are commonly found in the lake-shore banana-coffee land use system where mixtures involving food crops, coffee, cocoa and vanilla are intercropped with trees such as *Albizia spp. Ficus spp* and fruit trees such as *Artocarpus heterophyllus Psidium guyajava, Mangifera indica, Persia americana, Passiflora edulis, Citrus spp, Carica papaya* and *Canarium schweinfurthii.* The food crops include banana (*Musa spp*) beans *(Phaseolus vulgaris)* cassava (*Manihot esculenta*) sweet potato (*Ipomea batatas*) taro (*Colocasis spp*) maize (*Zea mays*) tomatoes *(Lycopersicon esculentum)* red pepper (*Capsicum annuum*) pineapple *(Ananas comosus)* doodo (*Amaranthus candatus*) etc.

Food production is the primary function of most home gardens and it is almost continuous throughout the year. The intimate mixture of various agricultural crops and multipurpose trees meets the basic needs of the local population while the multistoried configuration and high species diversity of the home gardens help reduce the environmental deterioration commonly associated with monocultural production systems (Nair 1993).

A recent study in Mukono district (Nielsen *et al* 1995) revealed gender differences in preference for tree species found in home gardens. While women reportedly prefer fruit trees and firewood species, men's preference is for timber and construction. Therefore, attempts to improve the use of trees in home gardens should involve men, women and the children to better understand how they utilise trees and what products are considered priority.

Judging from their geographical distribution, home gardens appear to be sustainable; and efforts to improve small holder production of staple foods, such as banana, should aim at improving the management of home gardens. The following are some of the benefits of growing bananas under home garden settings.

(a) Reduced soil erosion: The multilayer plant cover protects the soil from direct impact of raindrops, while the dense root system binds the soil. Consequently, erosion is generally low under a home garden.

(b) Increased soil organic matter: The abundant litter and root system of home gardens increases organic matter content of the soil. Soil fertility is enhanced by the good decomposition rate of organic matter favoured by high soil moisture. Soil structure is also improved by abundant organic matter which increases porosity and aeration.

(c) Increased soil moisture: This is achieved by intercepting and holding water through litter and plant cover and by reduced runoff and erosion.

(d) Reduced soil temperature: Plant and litter cover reduce soil temperature by intercepting solar radiation. Water evaporation is reduced and daily variations of soil temperature are minimized. Low soil temperature is good for biological activity of soil flora and fauna.

(e) Increased interception of light: Light is utilised by several layers of vegetation, leading to an overall increase in biomass yield from home gardens. Although photosynthetically active radiation (PAR) decreases downwards in the canopy, shade tolerant bananas can still grow in the lower layers. Moreover, as temperature also decreases downwards in the canopy, the resulting light and temperature microclimate seems to be good for bananas, coffee and other understorey species.

(f) Multipurpose trees and shrubs on crop lands: It is common in many parts of Uganda to find a variety of locally-adapted multipurpose trees widely scattered over farm lands or according to some systematic patterns. In the lake shore banana-coffee land use system, the trees are usually fruit species such as *Artocarpus heterophyllus, Mangifera indica, Persia americana* and others. Shade trees such as *Albizia lebbeck, Albizia chinensis,* and *Ficus natalensis* are deliberately left to provide shade to coffee and bananas. In the east and north east (northern and eastern cereals-cotton-cattle system) the common species include *Vitellaria paradoxa* (shea-butter tree),*Mangifera indica* and *Ficus mucuso* (as a shade tree). Intercropping under scattered trees is the simplest, most common and popular form of agroforestry among small holder farmers in Uganda.

(g) Boundary planting, in which trees are in systematic line-planting along farm boundaries and within crop is also widely practised. The commonest species include *Markhamia lutea, Senna siamea, Senna spectabilis*, and others. Poles and firewood are the main products from these trees. *Grevillea robusta* has recently been found to be suitable for boundary planting, and is demanded widely throughout the country.

(iv) Soil and water conservation barrier hedges

In areas with steep slopes, trees and shrubs are planted on bunds and terraces with or without grass strips to stabilise terraces. Recent research in Kabale has demonstrated the value of *Calliandra calothyrsus* hedges in controlling water and soil runoff (Kakuru, 1993). As a technology, soil conservation hedges are undergoing wider dissemination in Kabale district and other areas with steep slopes.

(v) Woodlots

Farmers who have large pieces of land reserve portions for woodlots for the production of poles and/or fuelwood. The commonest species for woodlots is *Eucalyptus*. The species is usually established in swamps where other crops can not grow well. Some farmers interplant it with sugarcane and cocoyams which are water-logging tolerant crop. Woodlots are widely distributed in Uganda. In the West Nile cereals-cassava-tobacco system, *Eucalyptus* woodlots are established to meet fuelwood requirements for tobacco curing.

Silvopastoral systems
These include both intensively and extensively managed types.

(i) Intensively managed

(a)   Protein bank (cut-and-carry) fodder production

This practice is commonly found in the lakeshore banana-coffee land use system, mainly around the major towns. The fodder trees are grown in block configurations or along plot boundaries, or interplanted in existing fodder banks of elephant grass, or on contour bunds in association with grasses. The foliage is lopped and fed to animals that are kept in stalls (zero-grazed). In addition, crop residues are used intensively and the manure is used in banana gardens.

Recent studies in Burundi have recommended the planting of *Calliandra/tripsacum* and *Leucaena/tripsacum* hedges for fodder production on contour bunds (Akyeampong and Dzowela, in press). The inclusion of tree fodder in a grass-based diet has two major benefits. First is the increase in feed intake which results in improvement in animal performance. The second benefit of feeding tree fodder hinges on the quality and the agronomic value of the manure - the N concentration of the litter increases.

Small scale dairy producers in Kabale and Mbale districts are increasingly including *Calliandra calothyrsus* on the terrace bunds for fodder production.

(b)　Live fences of fodder trees and hedges
In this subsystem, the fodder trees grow to the size of fence posts around plots and the trees are lopped periodically for fodder. In Uganda, *Ficus natalensis* serves this role in the banana-coffee lakeshore land use system.

(ii) Extensively managed
The system has got a range of subtypes ranging from situations where cattle graze under *Eucalyptus* plantations or other shade trees to browsing systems where foliage (tender twigs, stems, leaves and pods) of standing trees are consumed. In the former, the role of trees is indirect, for example by providing shade to animals and promoting grass growth. The role of trees in browsing systems is more direct than in grazing systems. In Uganda, extensively managed silvopastoral systems are commonly found in Mubende, Kamuli, North Mukono, Nakasongola, Soroti, Kumi, Moroto, Kotido, Apac, Lira and parts of Mbarara where *Acacia spp* are scattered in communal rangelands. The browse trees, bushes and shrubs act as protein banks during dry seasons.

Agrosilvopastoral

The most common example of agrosilvopastoral systems are again the home gardens already described, with a large number of herbaceous and woody plants and livestock. The animals are kept in stalls and zero-grazing, tethered or grazed on communal areas depending on the availability of land. It is a common practice in most land use systems of Uganda.

Other systems

(i) Apiculture with trees (Entomoforestry)
In this system trees provide bee forage for honey production. It is common in the wooded savannas of Lira, Apac, Kumi, Soroti, Kamuli, Nakasongola, north Mukono, Mubende and parts of Mbarara.

(ii) Buffer-zone agroforestry
Agroforestry practices are increasingly being introduced into buffer zones around protected forest areas as an option which may reduce pressure on forest resources and improve the living standards of the people living around these protected areas. In Uganda, buffer-zone agroforestry is practised around the Bwindi impenetrable forest reserve and mount Elgon forest reserve.

Buffer zone management is aimed at protecting forest reserves from effects of human encroachment so as to maintain biodiversity within the ecosystem. It is a new forest management approach in Uganda and is currently being evaluated.

Agroforestry systems with potential in Uganda

(i) Improved tree fallows
This term refers to improved alternatives to the fallow phase of shifting cultivation. Improved tree fallows are an alternative to shifting cultivation in areas where the latter is no longer possible, due to population pressure.

Suitable species for improved tree fallows, are those trees and shrubs with soil improving qualities. Recent research findings have for example shown *Sesbania sesban* as invaluable species for these practices. Although not commonly practised in Uganda, improved tree fallows have a high potential in the densely populated areas where the situation demands a shorter fallow (e.g., the Lake Victoria coffee-banana land use systems and the Kigezi annual food montane land use system).

(ii) Alley cropping
This entails growing food crops between hedgerows of planted shrubs and trees usually and (preferably) leguminous species. The hedges are pruned periodically to provide biomass (green manure) and to prevent shading of the intercrops. The addition of green mulch has a favourable effect on the physical and chemical soil properties and hence crop productivity. Kang and Duguma (1985) showed that maize yields obtained using *Leucaena leucocephala* leaf materials produced in hedgerows planted four metres apart was the same as the yield obtained when 40kg N ha-1 was applied to the crop. Several limitations with alley cropping include its unsuitability in moisture-stressed conditions, shade effects caused by the hedgerows, as well as the reduction of land available for crop production, the additional labour required to maintain and prune the hedges, and the fact that farmers may opt to use the prunings as animal fodder rather than adding it to the soil.

Alley cropping is not a common agroforestry system in Uganda but it would be suitable in highland areas with steep slopes where hedgerows are established to check water and soil runoff. Alley cropping would also provide green manure.

Agroforestry and soil conservation

On a global basis, it is estimated that every year 10 million ha of land fall out of production due to soil degradation (FAO,1983). The most prevalent forms of soil degradation are nutrient depletion and soil erosion. Soil degradation is a major problem particularly in developing countries where effective erosion control measures and nutrient replenishment are seldom practised. In Uganda, for instance, many farmers cultivate the same pieces of land year in and year out without fallowing or applying fertilizers. The result of this has been a consistent decline of the biological potential of the soils in form of reduced yields. Several surveys conducted in Uganda report that reduced yields due to fertility is a major problem in the country today. Trees are known to improve and maintain soil fertility using agroforestry practices as a low input system for the farmers.

Trees and other perennial plants grow well in ferrallitic soils of Uganda. Most areas where these soils occur also receive sufficient rain for tree establishment if planted early enough at the onset of rains. Generally, ferrallitic soils are old and highly weathered. They have little or no mineral reserves. They tend to have low pH and CEC. Inherent fertility is low to medium requiring use of fertilizers for sustainable production. They depend largely on organic matter as the major source of plant nutrients (Tumuhairwe, 1986). Their physical properties are suitable for plant growth. Nonetheless, under rainy conditions they are vulnerable to leaching, illuviation, weathering and degradation.

Ferrallitic soils are usually deep which is a major requirement in tree establishment and growth. Most ferrallitic soil have a udic moisture regime (moist) almost the year round. They tend to have an isohyperthermic temperature regime (8-10°C) which is favourable. Generally, ferrallitic soils support a wide range of crops, trees and shrubs. Given that most farmers are resource constrained, fertilizers are seldom applied. Use of suitable agroforestry species can improve on conditions of these soils.

Vertisols, on the other hand, occur in a semi arid areas of Uganda. Establishing trees in these areas may be a little difficult.

Trees and soil improvement
Shifting cultivation provides evidence that trees improve soil conditions. Soils that develop under trees are fertile, with good structure, water holding capacity, and resistant to erosion. However, when trees are cleared for cultivation the soil loses these qualities.

Edwards *et al* (1990) asserted that agroforestry systems can impact significantly upon soil fertility. Felker (1978) reported that *Acacia albida* increased organic matter content, water holding capacity and nitrogen by 50-100% as compared to surrounding areas without the tree.

Aggarwal (1980) observed significant increases in N, P, and K in the top 30cm of soils under *Prosopis cineraria* and *Prosopis juliflora* as compared to open field in north west India.

Radwanski and Wickens (1981) reported increases in pH, organic carbon, N, TEB, CEC and base saturation under *Azadiracta indica* as compared to open field in northern Nigeria.

In Uganda, several PRAs have been conducted, especially in areas above 1000m asl. A survey conducted in Iganga (Wajja and Mbalule, 1992, unpublished report) revealed that a number of indigenous trees were reported by farmers to have positive influence on soil characteristics. These trees included *Ficus mucuso, Ficus natalensis, Ficus brachipoda, Albizia coriaria,* among others. Another survey conducted in Mukono (Nielsen *et al*, 1995) revealed similar observation. Furthermore, *Albizia chinensis* was also reported to have positive effects on soil. These species were found growing together with crops such as bananas, coffee beans, maize.

Farmers further reported that crop yields under the trees were much higher than away from the trees on similar soils. Therefore, there is evidence that these trees could be potential agroforestry species. In both studies some trees were reported to have negative effect on soils. These included. among others, *Senna spectablis, Senna siamea, Spathodea campanulata*. There were mixed reports on some trees, notably, *Milicia excelsa* and *Markhamia lutea*. While some farmers reported that they were negative on soils, others reported them as being positive and others thought they were neutral. Following these observations more studies are being initiated by the Forestry Research Institute and Makerere University to characterize the influences of these trees on soils.

How trees enhance crop yields

A fertile soil is one which can bring forth good crop yields given a favourable biophysical as well as a disease and pest free environment. For a soil to be fertile, its physical, chemical and biological properties need to be conducive for plant growth. By improving on the physical, chemical and biological status of soils, trees can enhance crop performance.

Physical influence

Trees reduce direct rain drop impact on soil, thereby preserving the soil structure. It is widely known that soils under a mature forest have best structure. Such soils have granular particles, porous, well aerated and drained. These soils do not experience erosion as compared to cultivated soils. However, when trees are cleared for cultivation these factors begin to decline steadily and erosion sets in. By incorporating trees in the farming system the rate of decline is substantially reduced.

Chemical influence

Kellman (1980) observed that in some trees top soil nutrient enrichment is achieved at the expense of the lower horizons. This is particularly so if nutrients are washed down from the top soil and accumulate in the lower horizon. Since trees root deeply, they are able to extract these nutrients and bring them back to the top soil through litter fall and decomposition. This mechanism is usually referred to as 'nutrient pumping'. In western Kenya, *Sesbania sesban* is being used to pump nitrogen from lower horizons back to the plough zone in improved fallows during the short rains. In the next long rains, trees are cut down and maize is planted. Records indicate that there is up to 30% yield increment under improved fallows as compared to natural fallow or continous cultivation.

In other cases, however, nutrient enrichment is a result of preferential retention of atmospheric nutrients. Leguminous trees in association with prokaryotic organisms fix atmospheric nitrogen and incorporate it into the farming system. Classic examples of these include *Calliandra calothyrsus, Sesbania sesban, Albizia chinensis, Leucaena diversifolia*. It is known that some trees are also able to trap sulphur from the air and incorporate it to the system. Carbon is yet another nutrient added to the system. Most trees have large leafy biomass and fix large amounts of carbon through photosynthesis. In some cases, both nutrient pumping and preferential nutrient retention are employed in fertility enhancement.

Most tropical soils depend on organic matter as the major source of plant nutrients. Soil organic matter is usually a good indicator of fertility status. Many trees produce large amounts of leafy biomass. Reddy (1992) reported that *Grevillea robusta*, a popular agroforestry species, produces substantial amounts of litter. Furthermore, laboratory analysis revealed that the leaves contained 1.18% N, 0.12% P and 0.5% K, quite sizeable amounts of nutrients. Upon decomposition these nutrients are deposited in the plough zone. *Grevillea robusta* originates from Australia and adapts to a wide range of climatic and soil conditions. It has been successfully established in many parts of Uganda. It is a fast growing tree that produces substantial woody biomass and its competition with crops is minimal.

By increasing the level of organic matter in the soil, trees go a long way in enhancing soil fertility. Organic matter binds mineral particles into granules responsible for loose friable condition of productive soils. It improves on the balance between fine, medium and large pores, thus maintaining a balance between the liquid and gaseous phases of the soil.

Organic matter increases the water holding capacity of the soil. It is also a source of major plant nutrients, including N, K, S and micro nutrients. Organic matter increases available P by blocking fixation sites on iron based complexes. Organic matter provides a better environment for biological nitrogen fixation and also reduces nutrient leaching.

Biological influence

Organic matter is the major source of energy for soil organisms. By increasing the level of organic matter in the soil, trees can enhance fauna activities. Some of the macro fauna are useful in improving the porosity of the soil as they dig tunnels into the soil. Others assist in physical break down of litter into fine particles that can easily be attacked by micro-organisms. The micro fauna take on various activities including decomposition of organic matter, and nitrogen fixation.

References

Aggarwal, R.K. 1980. Physico-chemical status of soils under 'khejri' (Prosopsis cinerania Linn). In H.S. Mann and S.K. Saxena (eds). Khejri (Prosopsis cinerania) in the Indian desert - its role in Agroforestry. Jodhpur, India: CAZRI, 32-37.

Djimde, M and D. Hoekstra 1988. agroforestry potentials for the land-use systems in the Bimodal Highlands of Eastern Africa: Uganda AFRENA reports series No. 4 ICRAF, Nairobi 53-55pp

Edwards, C.A., R. Lal, P. Madden, r.h. Miller and G. house (eds). 1990. Sustainable agricultural systems 9:132-138.

FAO 1983. Guidelines for control of soil degradation, FAO, Rome.

Fernandes, E.C.M. and Nair, P.K.R. 1986. An evaluation of the structure ndfunction of tropical homegardens.Agricultural Systems 21:279-310.

Kang, b.T. and duguma, B. 1985. Nitrogen movement in alley cropping systems. In: Kanga, B.T. and Van den Heide, J. (eds), Nitrogen in Farming Sytems in the Humid and Subhumid Tropics, pp 269-284. Institute of Soil Fertility haven, The Netherlands.

Keliman, M. 1980. Soi; enrichment by neotropical savanna trees. Journal of ecology 67:567-577.

Lundgren, b.O. 1982. The use of agroforestry to improve the productivity of converted tropical land. ICRAF Miscellaneous Papers. ICRAF, Nairobi, Kenya.

Lundgren, B.O. and Raintree, J.B. 1992. Sustained agroforestry. In: Nestel, B. (ed). agricultural Research for Development: Potentials and Challenges in Asia, pp 37-49. ISNAR, The Hague, the Netherlands.

Nair, P.K.R. (1993). An Introduction to Agroforestry. Kluwer Academic Publishers in Cooperation with International Centre for Research in Agroforestry (ICRAF).

Nielsen F., Y. Guinand, and J. Okorio 1995. Farmer participatory diagnostic research: Lakeshore banana-coffee land use systems of Uganda pp:61-65.

Radwanski, s.A. and G.E. Wickens 1981. vegetative fallows and potential value of the neem tree (Azaarachta indica) in the tropics. Economic Botany. 35:398-414.

Reddy, A.N.Y. 1992. Grevillea robusta in coffee plantations of Kartaka. Proc. Int. Workshop on Grevillea robusta. Nairobi, Kenya 1:35.

Torquebiau, E. 1992. Are tropical agroforestry homegardens sustainable? Agriculture, ecosystems and Environment. 41(2): 189-207.

Tumuhairwe, J.k. 1986. The problems of soil degradation, conservation and Agricultural production in Uganda. Proc. 8th AGM of the SSSEA. Kampala Uganda 56-70.

3
MANAGEMENT OF NATURAL FORESTS

D. M. Byabashaija, F. Kahembwe and P. Ndemere

Introduction

The current natural forests in Uganda are made up of gazetted central and local forest reserves. They are managed by the Forestry Department and Uganda Wildlife Authority. In addition to these reserved areas, there are large areas of forests and woodlands in the private and communally owned land. All of these forests and woodlands cover approximately 4.9 million hectares of which only 40% is under government ownership, control and protection. Table 3.1 gives the area distribution and ownership of tropical high forests.

Table 3.1: Distribution of Uganda's gazetted forests reserves by major vegetation types (ha)

	Government-owned Forest land		Private Forest land	Totals
	Central and Local Forest Reserves (FD & Local Autho-rities)	National Parks Wildlife Reserves, Controlled hunting areas (UWA)	Public, customary and Private land	
Tropical High forest	346,000	228,000	351,000	925,000
Savannah Woodlands	423,000	950,000	2,602,000	3,974,000
TOTAL	769,000	1,178,000	2,952,000	4,899,000
PERCENTAGE	**16%**	**24%**	**60%**	**100%**

Source: National biomass Study (FD) 1999.

Evolution of forest policy and legislation

The first legislation on forestry in Uganda were the various agreements made between the protectorate government and the respective native authorities like Buganda, Bunyoro, Ankole and Tooro. Control of the forests was vested in the protectorate government by these agreements.

The Buganda agreement of 1900 prohibited the cutting of forest produce without a permit on any land.

The Tooro agreement of 1900 and the Ankole agreement of 1901 provided that all waste and uncultivated land and all forests were to be considered the property of Her Majesty's government and the revenue therefrom was to be included in the general revenue of the Uganda protectorate (Morris 1965).

Nicholson (1928) who had spent time in Kenya and Uganda studying the forestry conditions in the two countries produced a report which led to the framing of the first forestry policy for Uganda (Forest Department 1951). This policy gave forestry in Uganda a clear set of objectives, namely to:

(i) retain under forests or afforest all areas of land, the retention of which under forestry is considered necessary on climatic or other indirect grounds.

(ii) meet with due regard to vested rights, such of the demands of the population of Uganda as cannot be met by individuals or local administration efforts.

(iii) advise individuals and local native administrations in all matters pertaining to arboriculture or forestry.

(iv) manage the state forests of Uganda so that they will give best financial returns on the capital investments in conformity with the three preceding objectives.

This policy, which was recommended by Nicholson, 1929, was accepted by the protectorate government and has remained in force till today with minor changes in subsequent revisions.

Its first clause laid a foundation for the major task of creating forest reserves which was a major feature of the Forest Department's history. In 1939, the policy was revised following the same lines laid down 10 years earlier. This time emphasis was given to proper management of the forest estate and technical training which had started in 1931 at Kityerera in Iganga district. Training of local staff appeared as a definite aim for the first time in the 1939 policy.

After the second world war, Uganda turned its attention to developing her forest resources in a much more vigorous way. A new policy was, therefore, desirable to take account of this. The amendment of the ordinance in 1938 permitted local authorities to constitute and manage forest resources.   Since many of the local authorities had become active in forest resources management, a new policy was required to address this issue and was made in 1948.

After independence, the new constitution in 1967 amalgamated central and local government services and gazetted all local forest reserves as central forest reserves. Subsequently gazetted, in 1968, all local authority forest reserves were taken over by the central government and local authorities were no longer authorised to engage in forestry activities except the maintenance of few minor village forests. In 1970, a revised policy was produced and approved by the Minister of Agriculture and Forestry.

The current forest policy was gazetted in 1988 and it aims at a careful balance between the satisfaction of domestic and export demand for forest products, protection and conservation of natural physical, biotic and genetic resources of the forest ecosystems.

The aims of the present forest policy, the basis of current forest management practices, (The Uganda Gazette, 15 January 1988) are to:

- maintain and safeguard enough forest estate so as to ensure that:
  (i) sufficient supply of timber, fuelwood, poles and other forest products are available in the long term for the needs of the country and where possible for export.

  (ii) water supplies and soils are protected, plants and animals (including endangered ones) are conserved in natural ecosystems, and forests are also available for amenity and recreation.

- manage the forest estate so as to optimise the direct and indirect benefits to the country by ensuring that;
  (i) the conversion of the forest resource into timber, charcoal, fuelwood, poles. pulp and paper, and other products is carried out efficiently;

  (ii) the forest estate is protected against encroachment, illegal tree cutting, pests, diseases and fires;

  (iii) the harvesting of timber, charcoal, fuelwood poles and other products apply appropriate silvicultural methods which ensure sustainable yields and preserve environmental services and biotic diversity;

  (iv) research is undertaken to improve seed sources for planting stock and the silvicultural and protection methods needed to regenerate the forest and increase its growth and yield. Research is also carried out into new and existing forests products, including tourism and education with the object of maximising their utilisation potential. Research is undertaken to monitor and promote the preservation of environmental services and conservation of biotic diversity.

- promote an understanding of forests and trees by;
  (i) establishing extension and research services aimed at helping farmers, organisations and individuals to grow their own trees for timber, fuel and poles and to encourage agroforestry practices;

  (ii) publicising the availability and suitability of various types of timber and wood products for domestic and industrial use and publicising the importance of environmental services provided by the forest;

  (iii) holding open days at regular intervals in all districts to demonstrate working techniques and bring attention to the positive benefits of forestry;

iv) promoting scientific research, environmental tourism education and related activities inside the forest estate.

Natural forest management practices

Forest management practices in Uganda are characterised by the type of land ownership and control. Forest management on government land is the responsibility of the Forest Department. However, all forests regardless of ownership, have to fulfil the tasks of national utility. In practice, there are three distinct patterns of land ownership in Uganda which have a bearing on the management of the forest resource.

(i) Forest reserves: The Forest Department has exclusive rights under the Forest Act.

(ii) Public land: The Forest Department is mainly concerned with reserved tree species.

(iii)Freehold, mailo land (private land): The Forest Department has almost no say other than providing advice on the management of the forest and tree resources.

Forest management activities in forest reserves are guided by the Forest Policy and empowered by the Forest Act. From the time of its inception, management of natural forests in forest reserves in the country has had two major objectives, namely to:

- produce economically maximum sustained yield of hardwood timber and other wood products (production); and

- maintain specimens of characteristic plants, animals and other biological communities in the forest (protection).

In the former, timber and a variety of other commodities are produced. In the latter, the emphasis is on the provision of services such as soil and watershed protection, conservation of genetic resources and other values of the indigenous flora and fauna rather than material commodities. In practice, the distinction between the production of commodities and the provision of services is not always clear or rigid. The distinction is mainly manifested in the official classifications and is given spatial expression in the form of zoning systems.

Production zone

The production zone consists of areas where intensive silviculture is practised to increase yields of the forest products, mainly timber. Activities include enrichment planting, encroachment re-planting, salvage operations, refining, liberation, direct regeneration and general protection. Concessions for timber harvesting are awarded in this zone. Harvesting of non-timber forest products such as rattan canes, gathering and tapping on a permit basis is also permissible here.

Early silvicultural practices in natural forests in the country were aimed at encouraging natural regeneration of the more valuable species such as the mahoganies. The initial efforts of natural regeneration were of limited success and enrichment planting trials had to be undertaken.

This embraced various measures for improving the percentage of marketable timber species while protecting other useful trees of various species.

The period from 1930 to 1960 was one of considerable achievement in the evolution of methods of regenerating the natural forests for production of timber and other products, a field in which Uganda came to be recognised as a leader among tropical countries (Hamilton, 1984). Studies in forest ecology were initiated by Eggeling in the 1930s forming the basis of management techniques later followed up by Dawkins and others. The first research plots for determining the growth characteristics of natural forests under different silvicultural treatments were established in 1933 (Howard, 1991).

Large scale operations were initiated in Budongo and the Mengo forests in the late 1930s with enrichment planting at first carried out by line planting of nursery-raised seedlings of valuable tree species, such as *Entandrophragma*, *Khaya*, and *Maesopsis*, after selective harvesting of an initial timber crop.

According to Lamb (1968), 4452 ha. (11,000 acres) had been line planted in Budongo alone by 1950. Other work on small scale had been done in Mengo district besides trials in Semliki, Kibale and Bugoma. Nevertheless, the slow growth of the planted trees (mainly *Khaya anthotheca*) due to retention of much of the overhead shade, the severe destruction by wild game, especially elephants and the shortage of labour for weeding and game control led to a sudden stop in the expansion of enrichment planting in 1950.

The scheme was abandoned and replaced by planting of intensive compensatory plantations. This approach was also abandoned later in favour of the post exploitation treatment by use of a mixture of butyl esters of 2,4-D and 2,4,5-T in diesel oil. This mixture, commonly known as arboricide was used to eradicate undesirable species and open up the canopy to facilitate natural regeneration (Shelter wood system). However, the cost was prohibitive.

The refining treatment was considered dangerous to the ecosystem. The non-selective killing of many species that were not marketable by then was also viewed as myopic. There was an argument that such species could become economic as the country developed and more timber for various uses became necessary. This was particularly true of *Antiaris toxicaria* (Kirundu) a very tall tree with magnificent bole for which there was no market in the 1950's but which later found market in plywood manufacturing, box making and construction industry.

It was also realised that it would be advantageous and economical if undesirable species could be put to some use instead of poisoning them with the arboricide and leaving them to disintegrate, often with devastating effects on the developing regeneration. When the demand for charcoal allowed clearing of all except the desirables at no cost or even for some revenue to cover part or all the silvicultural operations, refining by charcoal burning became a standard practice particularly in the belts of natural forests of the northern shores of Lake Victoria. It was possible to continue planting in vacated kiln sites throughout the year except during severe dry spells.

Protection/conservation zone

The protection/conservation zones are managed purposely for:

(i)   Biological reserves; for conservation, in perpetuity, of the gene pool and to preserve its habitat in a viable state capable of being self sustaining.

(ii)   Germ-plasm store houses; for secure source of regeneration material capable of sustaining the productive capacity of the forest.

(iii)   Scientific repositories for studies of the development of diversity in biotic and edaphic components without human interference.

(iv)   Sites for tourism development; for ecologically sustainable development or non-destructive use of the forest such as conservation promotional developments like construction of tourism trails including tourist attractions and other tourist facilities.

(v)   Environmental protection units for climatic buffer, watershed protection, protection of downstream activities.

Recently, many forest reserves were uplifted to conservation status and gazetted as National Parks. This include the forests of Kibale, Bwindi, Rwenzori, Semliki and Mt Elgon.

Problems of forest management

There have been several contributing factors to problems in the management of forest resources, especially closed natural high forests in forest reserves mainly;

(i)   agricultural encroachment.

(ii)   illegal harvesting of wood and non-woody products.

(iii)   institutional deficiency in the country.

Agricultural encroachment

It is clear from Forest Department reports that illegal activities were not a serious problem before the 1960s (e.g., Forest Department, 1951; Forest Department, 1955).

However, since about 1972, there was considerable degradation of the Forest Department estate (Hamilton, 1984). Forests like Kibale, Mt. Elgon and Mabira experienced serious agricultural encroachment during the period 1961-1988. This was exacerbated by the political and economic turmoil that the country underwent and reached a climax during the period 1973 to 1983. This coincided with Land Reform Decree of 1975 which permitted any Ugandan to settle anywhere in the country, and the breakdown of law and order following the war in 1979.

Some local government officials looked at encroachers as a source of revenue to the districts and any eviction was a direct loss of revenue in terms of graduated tax and market dues. At the same time, schools, churches, health centres and other forms of infrastructure were constructed in encroached areas and evictions definitely would lead to much material loss.

Forest encroachment in the country was associated with areas of high population pressures such as the foothills of Mt Elgon and resettlement of immigrants like the Bakiga from Kabale, while in other areas it was associated with commercial farming like in Mabira forest. However, even in areas of low population density agricultural encroachment remained a big forest management problem. For example, the encroachment in Luwunga Forest Reserve (cultivation) and Kasambya Forest Reserve (grazing) in Mubende district. The district is not densely populated and has vast areas of idle arable land with elephant grass and spear grass growing.

In 1983, attempts to evict the encroachers from forest reserves were initiated but were sabotaged by the unfavourable political and security atmosphere prevailing at that time. In 1987, the eviction process was resumed and this time it succeeded. By 1990, all encroachers were evicted from the affected forest reserves. Some of the previously encroached forests have been re-stocked through enrichment planting supplemented by natural regeneration. Table 3.2 shows the extent of encroachment in some forest reserves (State of the Environment 1994.)

Table 3.2: Extent of encroachment on severely affected forest reserves by 1990

Forest reserve	District	Encroached area/ha
Mabira	Mukono	10,000
South Busoga	Iganga	6,000
Bukaleba	Iganga	4,500
Mt Elgon	Mbale & Kapchorwa	31,000
Kiboga	Kiboga	2,000 L
Bwezigola Gunga	Kiboga	3,500
Kibale	Kabarole	500
Nile Bank	Jinja	300
Kagoma	Jinja	300 L
Kisangi	Kasese	1,000

Source: State of the Environment for Uganda 1994.

Illegal harvesting of wood and non woody products
The utilisation/extraction of forest products albeit unauthorised and without any scientific study has been taking place in most forest reserves. Products commonly harvested include sand, poles, vines, palm leaves, barks of some trees and other medicinal plants. The worst of these has been illegal pitsawying where various categories of people have been involved including security personnel, local administration chiefs and even some government staff have been implicated.

In many cases, no selection of tree size is done, but random cutting regardless of the quality and quantity of sawn wood that can be obtained. Worse still, both old and young trees of valuable species are removed which greatly reduces the standing value of the remaining forests.

There are no proper data regarding the harvest of these products but it is acknowledged that a lot of forest-based products are illegally extracted for various uses.

Institutional deficiency in the country
For nearly 100 years, it has been the responsibility of the Forest Department to protect the gazetted forest areas and reserved tree species and apprehend culprits who contravene the Forest Act. The period up to early 1970s was a golden age for the Forest Department. A well developed management system for the gazetted forest areas existed in the 1950s and 1960s with controlled programmes of work. Extraction/logging was followed by enrichment planting. The forest and its products contributed substantially to the country's revenue through sales taxes, royalties and tourism. The initial success of the department could be attributed to the low population and abundant forest resources. The department however, was later frustrated by shortage of labour and resources as well as the growth in population and the increase in intensity of the demand for forest products.

The break down of government institutions in the 1970s and 80s resulted in large scale invasion of the forest reserves and other natural resources that had been previously managed by the state. The Forest Department like other public institutions in the country fell victim to greed, mismanagement and neglect that was rampant in those days.

The deterioration of the physical infrastructure of the department further impaired its ability to maintain the forest estate. Budgetary cuts, both in recurrent and capital expenditures negatively affected the operations of the department. This led to degradation of forest areas emanating from severe agricultural encroachment, illegal pitsawying, illegal charcoal burning, fuelwood collection and grazing. Millers felled immature trees and inappropriate logging/ extraction equipment degraded the forest areas, thereby damaging wildlife habitats and retarding natural regeneration. The management capability of the department came to its lowest ebb following the 1979 war when it lost many of its vehicles, basic tools and many of its houses country wide got destroyed.

In 1986, state institutions were strengthened and rehabilitated. Vast areas of forest estate which had been lost to encroachment were reclaimed and indisciplined harvesting of forests has been reduced and the forest estate is being put back to proper management. However, management of savanna woodland reserves is still inadequate because of limited resource inputs and these happen to be the main sources of fuelwood and charcoal.

Management of forests on private and public lands
Natural forests on private and public lands have never had any form of management plans and have continued to be utilised unsustainably. In accordance with the Forest Act, the Forest Department is responsible for tree species on public land classified as reserved forest produce to the government. However, in many cases, people who lease land from the Uganda Land Commission for agriculture do clear such land of almost all trees and shrubs.

The forest policy is rather silent about the management of natural forests on freehold or leased, land and yet these forests constitute a large portion of the total natural forest resource especially along the Lake Victoria shore belt. There is no legal authority for the Forest Department to institute any regulatory measures on these resources.

Other causes of forest degradation
Forests are always liable to suffer damage by natural forces such as fires, winds, drought, floods, insect, or animal damage and disease. The law itself can never prevent these things from happening but it can foresee the possibility that they may happen from time to time and make reasonable provision for preventive or remedial measures. Most of these damages by natural forces will be limited in extent if good silviculture is practised.

Fire
Fires are a seasonal occurrence in forests, occurring mainly during annual dry spells. Some fires are associated with poaching/hunting, some are started by trespassers, while others are started by honey harvesters.   Forest protection against fire is included in the Forest Act of 1964 which provides for;

(i)   the establishment and maintenance of forest reserves.

(ii) the authority to control and regulate human activities in the forests; and

(iii) the enforcement powers of the Forest Department staff.

In the 60s, fire protection measures were strictly observed almost in all forests. This included scuffled fire lines around the external boundaries outside which the vegetation was burnt early. Internally, all roads were cleared of vegetation. Fire towers connected to station offices by telephone were manned 24 hours during dry seasons. Hand tools, spray pumps and vehicles were available. Manpower for patrol, inspection and law enforcement was available.

At present vehicles, tools, equipment and sufficient manpower are no more. The only fire protection measure being practised is the external fire line and is also limited to a few forests. Fortunately, damage by fires has not been unduly great except in some of the drier areas of the country.

Game animals and vermin
Though natural forests in Uganda contain a very large variety of game animals and vermin, only a few big ones like elephants and buffaloes are responsible for extensive damage to forest trees. Animal damage to tree growth in natural forests may be divided into two main categories;

(i) Interference with regeneration and damage to young tree crops by trampling and
     grazing.
(ii) Damage to mature or semi-mature standing timber by bark stripping resulting in
     stem and butt rot.

Insect pests and fungal infections
Natural forests have excellent biological equilibrium and are more stable than plantation forests (Akanbi 1990). They are, therefore, less attacked and damaged by pests. Important pests include; the Mvule gall fly, *Phytolama* sp. which feeds on terminal buds and young leaves, resulting in gall formation. In severe infestations, stunting of affected trees occurs. No appropriate control measure is available but resistant species such as the West African Mvule *Milicia regia* exist. Introduction of such species was undertaken to reduce attacks from the gall fly.

The Mahogany shoot borer, *Hysiphilla robusta* attacks the terminal shoots of mahogany. It bores into the shoot, eventually leading to death of the infested shoot. No control measure is available, except relying on natural control factors. *Monocchamus centralis* bores into wood of living *Maesopsis* causing extensive damage. The affected tree dies or loses grade for timber. Control of this pest is dependent on natural factors. Another defoliator, *Nadiasa* sp. is a serious pest of *Maesopsis*. Only natural factors maintain this pest to low levels. Fungal diseases include Mahogany rot, *Fusarium* Canker on *Maesopsis emimil* and dieback.

Forestry research in Uganda

Forestry research has been part of the forest management activities since the inception of the Forestry Department early this century. The research function was basically the development of new methods and techniques to assist in forest management. The major thrusts of early research include:

- Development of management processes in logical series of steps (management plans) and techniques of sampling, harvesting, regeneration and tending techniques, particularly in the natural high forests;

- Growth and rotation ages of various species including growth stand curves and yields through a network of permanent sample plots in the various forests;

- Wood properties of various indigenous species and recommendations for appropriate utilisation;

- Description of various forest types including site classification, species evaluation and recommendation of suitable species for plantation forestry;

- Development of silvicultural techniques for the major plantation species mainly to increase productivity of both the pole and fuelwood, and timber plantations; and

- Protection of the forest crop from disease, pests, and fires right from establishment through growth to harvesting and protection of wood against attack from logging through storage and utilisation.

Since 1992 when NARO was established and took over the responsibility for forestry research from the Forest Department in 1993 FORI's objectives, were defined and include the development of appropriate technologies to:

(i) increase the productivity and supply of forest products and services on a sustainable basis in both natural and plantation forests;

(ii) ensure efficient and environmentally sound and socially acceptable forest management and utilisation systems, and

(iii) facilitate scientific forest-based conservation of biodiversity and environment protection. FORI's research activities are organised into three programmes namely:

Forest Management, Agroforestry, and Forest Products and Utilisation. The future plans are to generate technologies in all aspects of forestry, develop effective mechanisms for technology transfer so that the population of Uganda understand the "real value" of trees and forests and so contribute to their sustainable management.

4

PLANTATION FORESTRY

J. F. O. Esegu, G. A. Maiteki and P. Kiwuso

Planting of trees in Uganda is as old as agriculture since people have always planted both food crops and trees. Trees are grown for a variety of products and services. The establishment of industrial, pole and fuelwood plantations is, however, a recent development. The Forest Department records indicate that by 1908 some hardwood plantations had been established around government stations (Byarugaba 1994). These consisted of prime timber species indigenous to Uganda. The aim was to assess growth potentials of these species in order to increase their standing volume to maximise returns on investment. The species included *Milicia excelsa* (Mvule); *Entandrophragma utile, E. cylindricum, E. angolense,* and *Khaya anthotheca* (Mahoganies); *Markhamia lutea* (Nsambya) *Maesopsis eminii* (Musizi).

At the beginning of this century, a number of exotic tree species were introduced. In 1912, seeds of some exotic species, including *Pinus insignis, (P.kesiya) Eucalyptus, Cupressus spp. Cedrela toona* and *Black wattle, Acacia mollissima (A. mearnsii),* were imported and plants were raised and distributed for urban planting. In 1913, some 452gms of Black Wattle Seed were sent to Kigezi, and by 1935 there were very many trees of this species in the district with natural regeneration spreading outside the plantations. The bark was collected and exported for tannin production and the wood was locally popular for fuelwood and charcoal making. Interest in plantation forestry has continued due to two major factors; the factors favouring plantation development and problems with the natural forest resource (Evans, 1992).

(i) Factors favouring plantation development

In Uganda, like many other tropical countries, there are silvicultural, economic, social and environmental benefits which make plantation forestry attractive. The most striking advantage is rapid growth both in relative volume and value. This is achieved through careful pre-planting surveys which help match species with site conditions for optimum growth; development of stands of selected species which suit the needs of users; full stocking on sites to ensure efficient land use; and manipulation of spacing, thinning regimes and rotation length to produce desired mix of timber sizes, poles, pulpwood and sawntimber.

Plantation forestry can significantly aid economic development. The role of plantation forestry to development and industrial importance has been described by Brown (1967). Industrial plantation establishment as well as tree planting for social and environmental objectives are central to forestry development strategies in Uganda. The benefits of plantation forestry to the economies of Uganda and other developing countries includes the following (FAO 1985):

- Resource creation rather than sole exploitation to meet demand for wood and wood products.
- Use of land often of little or no agricultural value
- Development of infrastructure such as power transmission, communication services, houses, shops and schools often in remote areas.
- Integration of tree planting with other land uses to contribute to the environmental and other benefits from trees and forests.

The major environmental protection values for tree planting include soil stabilization, prevention of soil erosion, controlling water runoff in catchment areas, providing shelter from wind, heat, and dust storms. All plantation forestry developments have the carbon sequestration benefits, particularly when established in low carbon sink ecosystems such as grasslands.

(ii) Deforestation and degradation of natural forest resource

For a long time natural forests have been exploited, cleared, and degraded by man resulting in gradual decline in extent. Provisional estimates from surveys covering 62 countries (FAO 1990) indicate that tropical deforestation amounts to 16.8 million hectares per year. The 1992 Uganda National Environment Action Plan and FAO publications reported estimated annual deforestation rates of $500km^2$ and $650km^2$ respectively (NEAP, 1992). The main causes of forest destruction in Uganda are well known and include the following:

- forest clearance for agriculture;
- intensive logging for sawntimber;
- exploitation for charcoal, firewood, poles and other domestic uses;
- industrial expansion;
- war (civil strife)

Some areas of natural forest cannot be exploited because of physical limitations and fragility. For example, the steep mountainous terrains of Mts Elgon and Rwenzori and swampy areas of southern parts of Uganda. Such forests are needed for protection of such sites and ecosystems. Many forests in such areas do not have good infrastructure of roads and services. In addition some of these forests do not contain high volumes of utilisable timber.

The need to manage natural forests on a sustainable basis has proved difficult due mainly, among other reasons, to the lack of satisfactory natural regeneration (Poore, 1989). However, successful experimental trials to enhance regeneration in natural forests in Uganda have been undertaken (Dawkins 1958). The potential exists, but there is paramount need to relieve pressure on the dwindling natural forests by establishing plantation forests (Westoby, 1989).

Establishment and management of plantations

As early as 1900, the natural forest estate in Uganda was considered to be inadequate for sustainable supply of wood products and services. Troup (1922), recommended afforestation schemes to extend the country's forest cover. This was subsequently emphasised in the forest policies of 1929 and 1948.

In order to support the development of plantation forestry, research was undertaken to select fast growing species suited to various sites in the country. Species research to select suitable species for planting in Uganda started in 1908. It was executed locally at first by the district forestry staff and the local government authorities. The research was greatly extended after 1953 when methods of research planning and recording were established in the first written silvicultural research plan, (Dawkins 1953).

Data from about 6500 active research plots were appraised and summarized in 22 technical reports 'Technical Notes of the Forest Department' (1968). The classification of research plots was described by Stuart-Smith (1967). The performance of individual species and recommendations were summarised by Kriek (1970). In addition to species evaluation, techniques for raising, establishment and management of the plantation crop were developed (Karani 1978).

(i) Conifer timber plantations

The species identified for timber plantations in Uganda include *Pinus caribaea* var. *hondurensis* and *P. oocarpa* in the savannah areas and dry secondary forest at low and medium elevations of between 800 to 1200 metres above sea level. At medium and high elevations above 1200 metres, *Pinus patula, P. kesiya* and *P. caribaea* var *bahemensis* and var *caribaea* are recommended, with *Cupressus lusitanica* where there are deep rich soils.

Softwood timber or coniferous forest plantations started at Muko Forest Reserve in 1942. It continued to cover the forest reserves of Lendu, Mafuga, Mwenge and increased all over the country with support from the Norwegian Agency for Development Cooperation (NORAD) to its peak in 1971. The planting target was 2000 ha per year, (1971-1976 Five year Development Plan). However, after 1978 the planting programme declined and by 1982 it had virtually stopped. The total area planted had reached about 14,000 ha. Since then a small area has been planted and this has brought the total plantation forests to approximately 15,000 ha. This is far below the target of 50,000 ha planned as an ideal supplement to the natural forest by the year 2000. The area distribution of softwood plantations by district is shown in Table 4.1.

Table 4.1: Area distribution by species of softwood timber plantations in Uganda 1997

District	Forest reserve	Species planted area (ha)			Total area planted (ha)
		P.caribaea/ oocarpa	*P.patul*	*C.lusitanica*	
Arua	Okavureru	420			420
Gulu	Abera & Opit	612			612
Hoima	Wampanga	518			518
Iganga	Namafuma	108	6		114
Iganga	Bukaleba	168			168
Jinja	Namasiga	360			360
Kabale	Mafuga		1071	537	1608
"	Muko		31	56	87
"	Kiriima		528	96	624
Kabarole	Kanyawara		198		198
"	Kagorra		215	499	714
"	Kikumiro		474	256	730
"	Kyehara	235	248		483
"	Oruha	347			347
Kapchorwa	Kapkwata		656		656
"	Suam			426	426
Kiboga	Nakwaya	477			477
"	Kikonda	296			296
"	Zimwa	55			55
"	Lukuga	112			112
Kitgum	Lututuru	80			80
Lira	Kachung	284			284
Luwero	Katugo	1793			1793
Masindi	Nyabyeya	200			200
Mbarara	Bugamba	638	572		1210
"	Rwoho	283	488		771
Nebbi	Usi		433		433
"	Awang		163		163
"	Lendu	44	1229		1273
Soroti	Kateta	47			47
"	Pingire	160			160
TOTAL		**6486**	**5656**	**1870**	**15413**

Source: Forest Department Records 1997

(ii) Management of softwood timber plantations

Many different operations are needed right from producing suitable seedlings for planting through establishment care or tending to harvesting and regeneration of the plantation crop. As a plantation is a growing resource, its management is very important so that appropriate and timely interventions are made to modify both the quality and quantity of the crop and end-products. The plantation life history of silvicultural operations, management options and decisions to be taken are shown schematically in Table 4.2.

The management alternatives which directly affect the quality of end-products are species choice, including seed source; initial spacing; pruning; thinning, and rotation length. Obtaining information regularly about resources growth, yield, pest and disease problems in forest plantations provides the essential data for proper management.

The initial spacing for conifer timber plantations in Uganda is 2.74 x 2.74m and planting holes are dug about 15cm deep and wide. Planting is carried out in April in all parts of the country except for the South-western areas where the long rains occur during September to December and planting is done during October-November. Good soil moisture build-up based on rainfall reliability is essential for subsequent high survival of seedlings. Three months after planting, stock checking is undertaken to assess the survival and seedlings requirements for beating up, which is carried out during the following rain season, or one year later, when the soil moisture build-up is sufficient. Weeding, i.e., slashing for pines and spot hoeing for cypress and hardwoods is undertaken to reduce competition from other vegetation and prevent damage from rodents.

In each management plan, area pruning and thinning schedules have been developed. They are normally based on local growth rates of individual species and envisioned market requirements. Table 4.3 gives the guide to pruning and thinning in conifer plantations for industrial timber production.

- Where rodents are a problem, 'rodent pruning' is done at age one to two (1-2) year on all trees/plants.

Table 4.2: Schematic representation of plantation life-history

Silvicultural Operation	Main decision	Crop description (trees & stand)	Time from planting in years
Obtain seed	Species Provenance		-2 to -1
Raise plants in nursery	Bare rooted or container plants	Seedlings	-1
Prepare ground	Site preparation and rate of planting		-0.5 to -0.1
Planting	Spacing	Forest plantation seedlings	0
Tending	Weeding	Saplings Canopy closure	0 -3
Low Pruning	Level of pruning	Small trees/small poles	4 -10
Thinning High Pruning	Thinning and pruning frequencies and intensities	Big trees/large poles immature stand	5 -15
Clear felling	Rotation age	Mature trees Final crop	20 -25
Replanting	Same species or new	Second rotation	

Table 4.3: Guide to thinning and pruning in Uganda plantations

Age in years	Operation	Thinning (T) (tree left per ha)	Pruning (P) (Maximum half height)		Remarks
			metres	No to prune	
1		1370			Planted at 2.7x2.7m
5 - 7	P.1		2	All save runts (stunted growth)	Access-year Before T1
6 - 8	T.1	990+ 10%			
7 - 9	P.2		5	All	Year after T1
11 - 13	P.3		7	300-345	
13 - 15	T.2	570 -640			Occasionally down to 490
14 - 16	P.4		10	300-345	
20 - 22	T.3	300-345			

Note: In 1969, it was recommended to limit pruning height to five metres (5 m), equivalent to one bottom log on each tree. The argument was that the cost of pruning to 10 m was higher than the price paid for the logs.

(iii) Pole and fuelwood plantation
The most productive species for fuelwood and poles in Uganda are *Eucalyptus* species. *Eucalyptus camaldulensis*, and *E.citriodora* are suited to the dry northern parts of the country. *E.tereticornis* is suitable for the central and medium rainfall parts of the country while *E.grandis* grows very well in the Lake Victoria region and at medium to high altitude areas of the country, yielding both poles and sawn timber.

During the 1930s, *Eucalyptus* species were established around urban areas in the country. These plantations continued to expand and by 1946 all the major towns were served with pole and fuelwood plantations. Subsequently, plantation management weakened due to both political and economic decline and most of the pole and fuelwood plantations became degraded.

The multi-donor Forestry Rehabilitation Project (1988-1996) included in its programme the rehabilitation of some peri-urban plantations. This was under the peri-urban and Integrated Pilot Wood Farms Project funded by NORAD and implemented through the Norwegian Forestry Society in co-operation with the Forest Department. The project has continued with the rehabilitation of the peri-urban plantations through allocation of plots to individuals and groups in order to increase the supply of fuelwood and poles to meet the growing wood deficit. The current status of the peri-urban plantation areas for poles and fuelwood production in the country is shown in Table 4.4.

In addition to the Forest Department plantations, there are other fuelwood and pole plantations under various agencies including sugar, tea, tobacco estates, tree planting groups and individuals.

(iv) Establishment and management of poles and fuelwood plantations
The basic management schemes for pole and fuelwood plantations are similar to softwood conifer timber plantations (see Table 4.2).

The normal spacing for *Eucalyptus* plantations in Uganda is 2.4 x 2.4m for fuelwood crops on 4 years rotation on good sites and good management, (5-6 years on poor sites), and 1.8 x 1.8m for pole crops. Clean hoeing in grassland areas is essential for the establishment of new plantations until canopy closure. Coppice crops do not require hoeing but slashing back of competing weeds, climbers and grass. Coppice management requires reduction to 2-4 stems per stool when the plants are 3-5m tall. Felling should be done as low as possible and stumps must be smooth and sloping. Felling should be done by blocks of no bigger than one hectare and must be completed as quickly as possible. All stems must be removed before another block is started. Coppice regeneration after haphazard or selection felling is usually poor.

Table 4.4: Poles and fuelwood peri-urban plantation area in Uganda by 1994.

District	Forest Reserve Area (Ha)	Area Planted By FD	Area Planted By Out-growers	Total Planted	Un-planted
Apac	100				100
Arua	638	313	148	461	177
Entebbe	289				289
Gulu	114	50		50	64
Jinja	2730	375	1710	2085	645
Kampala	810	140	100	240	570
Kitgum	21				21
Lira	290				290
Mbale	562	326	104	430	132
Mbarara	210	55	20	75	135
Mbarara (Kyahi)	4000	260	204	464	3536
Masaka	410				410
Rakai	120				120
Soroti	398	17		17	381
TOTAL	**10692**	**1536**	**2286**	**3822**	**3870**

Pests of plantation forests

Plantation forests in Uganda consist largely of exotic tree species which are susceptible to a number of pests. The pest categories include wood borers, sapsuckers, defoliators and fungal diseases.

(i) Wood Borers

The most important wood borers are termites which generally attack all exotic tree species, especially during the early stages of establishment. Young trees and seedlings are particularly attacked by **Macrotermes,** but older trees can also be attacked by **Coptotermes** and occasionally by **Odontotermes**. Damage in young eucalyptus establishments can be so severe that complete replanting may be required, especially in the dry savannah areas.

Termite control in plantation forests is largely dependent on the use of chemicals. Use of botanicals, resistant and or repellent tree species, and biological control, though occasionally referred to, has not been evaluated. For chemicals, organochlorines such as Aldrin and Dieldrin were previously used. However, these were banned due to their high toxicity and persistence in the environment. New chemicals such as Marshal-Suscon (Carbonsulfan) have been introduced and are currently being used. Application of the insecticide at the nursery stage can protect trees from termite attack for up to two years.

Other wood borers include *Oemidia gahani* on cypress, and the long horned beetle *Analeptes triafasciata* on eucalyptus. These are, however, occasional in nature.

(ii) Sapsuckers

These largely consist of exotic conifer aphids accidentally introduced from outside the African continent. They include: the cypress aphid, *Cinara cupressi Buckton,* the pine woolly aphid, *Pineus(?) boeneri* and the pine needle aphid, *Eulachnus rileyi Williams*. The most important of these is the cypress aphid (Fig.4.1) which has caused extensive damage to cypress plantations, especially in south western Uganda. Heavy populations of the cypress aphid cause branch die back and tree mortality (Fig 4.2). Sooty moulds associated with *Cinara* colonies cause foliage discoloration and interfere with photosynthesis and gas exchange (Ciesla 1991).

The pine woolly aphid causes some considerable damage in pine establishments. The aphid feeds at the base of the needles and on the young bark. All stages of *P.boeneri* except the egg and first instar produce a white woolly waxy material which gives the species its common name (Mutitu at al. 1994). It is the white wool, (Fig. 4.3) rather than the insects themselves, that is commonly seen in the field. The insects are small and difficult to see with a naked eye.

The pine needle aphid attacks pine but its damage to trees is insignificant. All stages feed on the underside of pine needles of all age classes. Infested needles turn yellow and can be covered with sooty mould produced by the aphids.

Control of the cypress aphid using chemicals has been tried on live hedges (Kiwuso 1991). Lambdacyhalothrin, (Karate), Deltamethrin and fenitrothion were found to be effective. However, their use in cypress plantations is not recommended. This is because they are prohibitively costly, lead to environmental pollution and offer only temporary control.

Plate 4.1: Cypress aphid on a cypress branch in Mafuga, Kabale

Plate 4.2: Branch die back caused by Cypress aphid infestation and feeding, Mafuga, Kabale

Plate 4.3: White wool produced by the Pine woolly aphid on a branch of pine at Kirima, Kabale

Plate 4.4: *Gonometa podocarpi* feeding on Cypress at Muko, Kabale

Biological control as an option has been initiated using a parasitoid, *Pauesia juniperorum* against the cypress aphid, and a predator, *Tetraphlaps raoi* against the pine woolly aphid. Work to determine their efficacy is still under way.

(iii) Defoliators

These are occasional pests occurring mainly in conifer plantations. Most common ones include *Gonometa podocarpi*, Fig. 4.4, *Buzura edwardisi* and *Pachymetana sanguicineta*. Control of these is largely dependent on natural factors.

(iv) Fungal Diseases

These are important during all stages of plantation establishments. Damping-off and root rot cause considerable damage in nurseries. Damping off is caused by species of fungi in the genera *Pythium, Phytopthora, Fusarium, Rhizoctonia* and *Sclerotium*, acting singly or together (de Guzman et al. 1990). Proper drainage in nurseries can reduce the disease incidence and severity. *Dothistroma pini* is a serious disease of young or mature *Pinus radiata* and has prevented the establishment of this tree species in Uganda. Planting of resistant species, such as *Pinus patula and Pinus caribea* was recommended, this reduced the incidence of the disease.

(v) Vermin

Rodents, particularly the giant rat, can cause tree damage in young plantations. Slashing the undergrowth and use of appropriate rodenticides can be used to reduce damage.

Future for plantation forestry

The vision for plantation forestry in Uganda is to develop an efficient plantation resource, secure, well managed and economically productive to provide raw material supplies for forest industries on all scales. There are over 100,000 ha readily available to establish new timber plantations in forest reserves and much more outside the forest reserves. Substantial investment in commercial plantation forestry has started with individuals, groups, and various organisations who are leased unplanted forest reserves. The Forest Department is stepping up the plantation forestry programme in reserved savannah woodlands and grasslands. There are also increased tree planting activities by the public establishing woodlots and various tree arrangements for both products and services of trees in the environment.

5

FOREST PRODUCTS AND UTILISATION

P. J. Turyareeba

Forests have often been viewed as a source of removable products like timber, poles, firewood and charcoal only. However, forests also provide non-wood forest products (previously referred to as minor forest products). Non-wood forest products include fibres such as leaves, cane and rattan; food such as fruits, nuts, vegetables and fodder; and, extractives such as oils, gums, resins and pharmaceuticals for medicines or pest repellents. Forests also provide a number of services which include:

- habitats for wild life;
- soil management through fertility maintenance and rehabilitation (chemical and physical);
- water management by water absorption and/or retention (in soil or vegetation), improved drainage and flood control;
- wind shelter through shelterbelts and windbreaks;
- contributing to the hydrological cycle through evapotranspiration; and,
- acting as a "carbon sink" for green house gas abatement through photosynthesis.

Forests and trees are a renewable natural resource which can be effectively and efficiently managed using appropriate technologies for sustainable socio-economic development and environment protection. This chapter will concentrate on wood and non-wood forest products.

Wood Fuels

Energy is a dynamic force in the growth and development of human society. The demand for energy is not an end in itself but for services like illumination, heating, cooking and mechanical power. In Uganda, biomass energy (charcoal, fuelwood and plant residues) accounts for 96% of the country's energy consumption as shown in Table 5.1. Of the estimated total annual wood fuel supply, 30% is obtained from bushland, 20% from woodlands, 48% from arable agricultural land and fallow land and only 2% from tropical high forests (UNDP/ESMAP, 1996). Wood fuels are used in the household, commercial and agro-industrial sectors.

Table 5.1: Uganda's fuel energy consumption

Energy Source Sector	Biomass	Petroleum	Electricity	Total
Household	86.9%	0.3%	0.6%	87.8%
Industrial	4.6%	0.6%	0.8%	6.0%
Commercial	2.8%	0.3%	0.4%	3.5%
Institutional	2.2%	0.1%	0.2%	2.5%
Transport	0.0%	0.2%	0.0%	0.2%
Total	96.5%	1.5%	2.0%	100.0%

Source: UNDP/ESMAP, 1996

Fuelwood

Fuelwood is the major source of energy in Uganda. However, its contribution to the national economy is often overlooked, depicted in the limited resources allocated to development of the biomass energy sector. Fuelwood is used in agro-industries (tobacco, tea, sugar and jaggery processing); construction (brick and tile manufacturing, lime production); and the food and beverages industry (fish smoking, food drying, bakery, alcohol and local brewing). Over 1.3 million tonnes of wood were used in industry in 1994, 50% of which was used by the lime industry (Anon 1995).

In the household and commercial sectors (hotels, restaurants, institutions) fuelwood is the major source of energy for cooking. The main technology used is the three stone hearth, a stone age tool which is inefficient, using only 10% of the energy available in the wood (Turyareeba, 1990).

Fuelwood is obtained mainly from public land, private land and to a limited extent from forests. In most cases, wood is collected from naturally growing trees. A number of large consumers have attempted to establish their own woodlots. There are over 20,000 ha of private *Eucalyptus* spp. plantations, most of which have been established by the tobacco and tea industry (LTS, 1995). In addition, there are a large number of small eucalyptus woodlots, particularly in districts such as Masaka and Kabale where wood fuels are scarce. The Forest Department has over 4,000 ha of plantations, out of which about 30% is managed using an out-growers scheme, for provision of poles, posts and firewood (Anon 1997a). These plantations were established outside major urban centres.

Fuelwood collection for household use is no longer considered to be a major cause of deforestation since it has been established that only twigs and dead trees are harvested for this purpose. Wood fuel harvesting for the commercial sector, agro-industrial sector and charcoal production contribute to environmental degradation and reduce carbon sinks since whole trees are often cut down for these end-uses.

Attempts have been made to reduce firewood consumption in the commercial and household sectors by introducing improved stoves. Non-governmental organisations and the private sector have taken a leading role in disseminating these technologies while the Forest Research Institute provides technical assistance. A large number of improved wood burning stoves have been installed in schools, colleges, prisons and hospitals. In addition, a few restaurants and hotels have invested in improved wood burning stoves. It is difficult to estimate the number of household stoves installed in the country since an end-user built approach is used in wood stove dissemination.

There is growing interest in woodlot establishment throughout the country. While most of the woodlots are meant for provision of poles, the tops and branches are used for fuel.

The use of wood for cooking on the three stone hearth particularly in the household sector takes a great toll on the lives of women and children who are responsible for collecting firewood. Only 10% of the energy in the fuelwood is used for the cooking task. The rest of the energy is lost by evaporation, conduction and convection. Furthermore, women and children are exposed to pollutants which have adverse effects on their health. Although no studies have been conducted in Uganda, it is estimated that cooking on the three stone hearth using firewood exposes women and children to large quantities of pollutants, above the WHO recommended levels of exposure (Hong, 1995).

Charcoal

Charcoal, a derivative of fuelwood, is an important household fuel and to a lesser extent, industrial fuel. It is estimated that 388,000 tons of charcoal were used in 1994, of which 86 % was used in the household sector (UNDP/ESMAP, 1996). Charcoal production generates over UShs. 16 billion annually, out of which 25% is collected as local authority or Government taxes and revenue (UNDP/ESMAP, 1996).

Charcoal is used mainly for cooking in urban and peri-urban households; and, in the commercial sector (mainly restaurants) for cooking and roasting meat. Charcoal is also used in local foundry. It is made in rural areas in Uganda. The major source of wood is public and private land, and to a lesser extent forest reserves. Landlords enter into agreement with charcoaliers who clear the land and convert wood into charcoal. However, this applies to only 20% of all land cleared for agriculture and livestock (Anon, 1995). The rest of the land is cleared using the slash and burn method which not only wastes the resource but also contributes to emission of greenhouse gases.

Charcoal production involves burning wood with limited supply of air. The earth kiln is used to make charcoal in most parts of Uganda. This involves felling and cross-cutting trees into billets. The billets are stacked and covered with grass (Plates 5.1, 5.2, 5.3 and 5.4). The kiln is covered with earth before it is lit. The kiln is monitored for a number of days (depending on the volume of billets used), after which the earth is removed. While removing the charcoal, charcoaliers run the risk of burns since at times there are glowing coals in the kiln.

Over 12,000 people are self-employed charcoaliers. About 60% of all charcoaliers are full time producers, half of whom are second or third generation charcoaliers (Anon, 1995). This implies that a mass of skilled producers has emerged over the years. Charcoal production has generated many more urban jobs for distribution and marketing.

Charcoal producers obtain licences from the Forest Department at a rate of U Shs. 9,600/= per charcoalier per month. Between January and June 1997, the Forest Department raised about U Shs. 15,000,000/= from 19 districts, equivalent to 13 % of expected monthly revenue (Anon. 1997b). This implies that over 90 % of all charcoaliers operate illegally. Even those who pay for licenses do not pay on a monthly basis.

The Forest Research Centre (now FORRI) carried out a number of studies in the 1960s, 1970s and 1980s which led to the development of the Mark V steel kiln and the Katugo (Missouri) kiln. The Mark V steel kiln was used in the late 1960s by charcoaliers but was later abandoned when the first kilns which were obtained on loan wore out. The Katugo kiln has never been used by charcoaliers (Turyareeba, 1991). The cost of the technologies developed appears to be a major limiting factor in technology adoption. In addition, the technology dissemination agency was also the policing agency, thus causing animosity between the technology disseminator and the end-user. Furthermore, charcoaliers in Uganda operate individually or in very small groups of two or three, thus making it difficult for the dissemination of improved charcoal making techniques and technologies.

Charcoal is marketed by the private sector. A large number of lorries and pick-ups deliver charcoal to urban centres daily. About half of the national charcoal demand is used in Kampala mainly supplied from Luwero, Mpigi and Mukono districts (Anon. 1992). Charcoal transporters are expected to obtain movement permits from district authorities before they can transport charcoal from rural areas to urban centres.

While demand for charcoal in urban areas continues to increase, the charcoalier has to work hard to meet this demand. Charcoal production is considered to be a major contributor to deforestation (Karekezi and Turyareeba, 1994). It is believed that charcoal production efficiency is still as low as 10%, efficiency obtained in studies conducted in the early 1960s (Mabonga-Mwisaka, 1972). This may not be the case, since research studies carried out in other African countries have shown surprisingly high efficiency for the traditional earth kiln under field conditions (Hibajene, 1994; Boiling Point, 1991;).

Efforts have been made to disseminate improved charcoal stoves in the urban centres. These stoves reduce charcoal consumption by 20 to 35 %. However, only 15 % of the households in Kampala and Jinja have bought these stoves (UNDP/ESMAP, 1996). Production of improved charcoal stoves has created employment for semi-skilled workers in the urban areas. Hawkers are often seen carrying three to four stoves in Kampala. This marketing technique has helped to increase the numbers of stoves in use.

Use of charcoal stoves indoors results in emission of carbon monoxide. Carbon monoxide even in low concentrations is a very potent poison mainly because it interferes with the oxygen-carrying capacity of the blood and therefore deprives body tissues of oxygen. Symptoms of acute carbon monoxide poisoning are headaches, drowsiness and loss of consciousness (WHO, 1992). Prolonged exposure may lead to physiological disturbances such as reduced blood pH and reduced birth weights in infants. Women are at greater risk of CO poisoning because they have lower levels of haemoglobin hence more predisposed to anaemia. Carbon monoxide is especially dangerous to foetuses because they mainly rely on their mothers to fulfil their oxygen demand through the placenta (WHO, 1992).

Timber

Trees are felled and milled to provide timber and logs for furniture and construction; and for veneer and plywood respectively. Hand tools, carvings, musical instruments, sporting goods, utensils, weapons (bows and arrows; gun butts, batons), wheels/spokes are made from timber, branches or logs. Wood for the above mentioned products is harvested mainly by private sector companies and individuals using simple technology such as hand axes, pangas and saws for felling and hand sawing; or, advanced technologies such as power saws for felling and mobile or stationary sawmills for conversion. Despite a large number of sawmills in the country (tables 5.3 & 5.4) 90 % of timber from natural forest is hand sawn while 95 % of the timber from plantations is machine milled (Lubega, 1997).

Timber and paper showed a 55.7% growth during the period between September 1995 and September 1996 (Ministry of Planning and Economic Development, 1997a). Total annual consumption of sawn wood is estimated at 100,000 - 150,000m³ (Kityo and Plumptre, 1997). Sawmills supply about 10% while the rest is hand sawn (Natvig, 1996). Timber is obtained mainly from forest reserves (tropical forests and plantations), public and private land. The annual allowable cut from natural forests, plantations and public/private land is estimated at 200,000m³, 150,000m³ and 100,000m³ respectively (Kityo, per. comm.; Carvalho and Pickles, 1994). Figures 5.1, 5.2 and 5.3 show timber from different species marketed for the three classes between June and December 1996. Demand for sawn timber is estimated to grow at 8% per annum (Kityo and Plumptree, 1997).

In the early 1990s 96% of all timber available on the market was obtained in natural forests out of which 65% were mahoganies (Anon.1994b; Khaukha, 1997). Currently about 6% of all timber entering the timber monitoring database is Mahogany obtained in the tropical forest, an appreciable improvement from the early 1990s (Anon. 1997b). However, it should be noted that about 40 % of the timber on the market enters the database, since there are a number of unregistered operators. Efforts to improve data collection are supported by the Nature Conservation Project.

The Forest Department encourages harvesting and use of timber from conifer plantations. During the 1980s there was hardly any conifer timber on the market. According to the Forest Department's timber monitoring database, 31% of the timber supply is from conifers.

Fig. 5.1: Class 1 Timber species marketed (June to December, 1996)

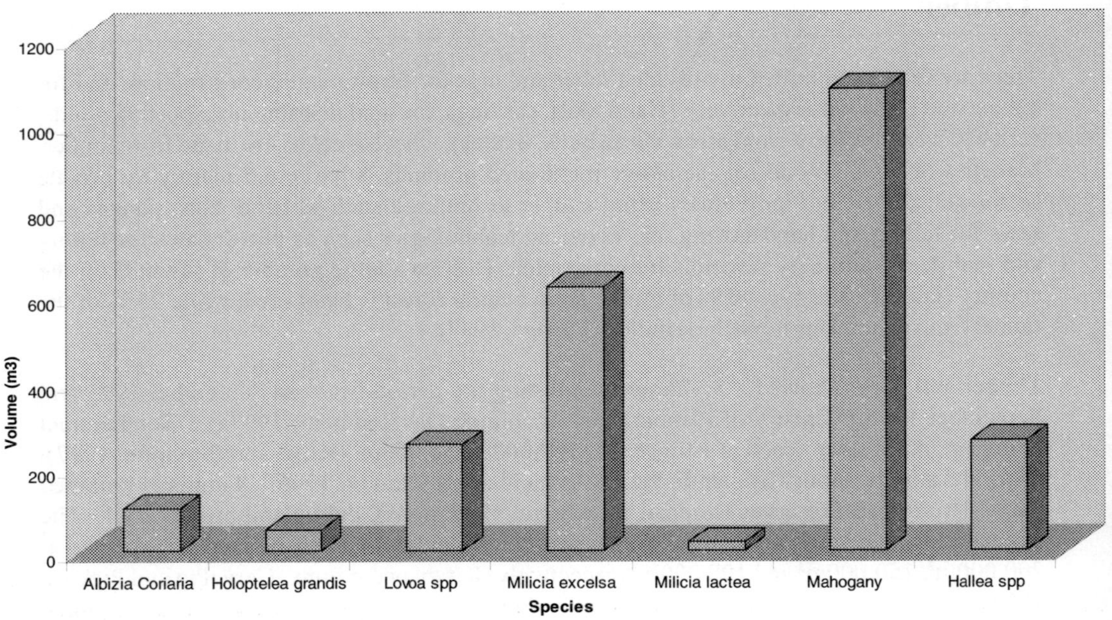

Source: Anon, 1997b

Fig. 5.2: Class 2 Timber species marketed (June to December, 1996)

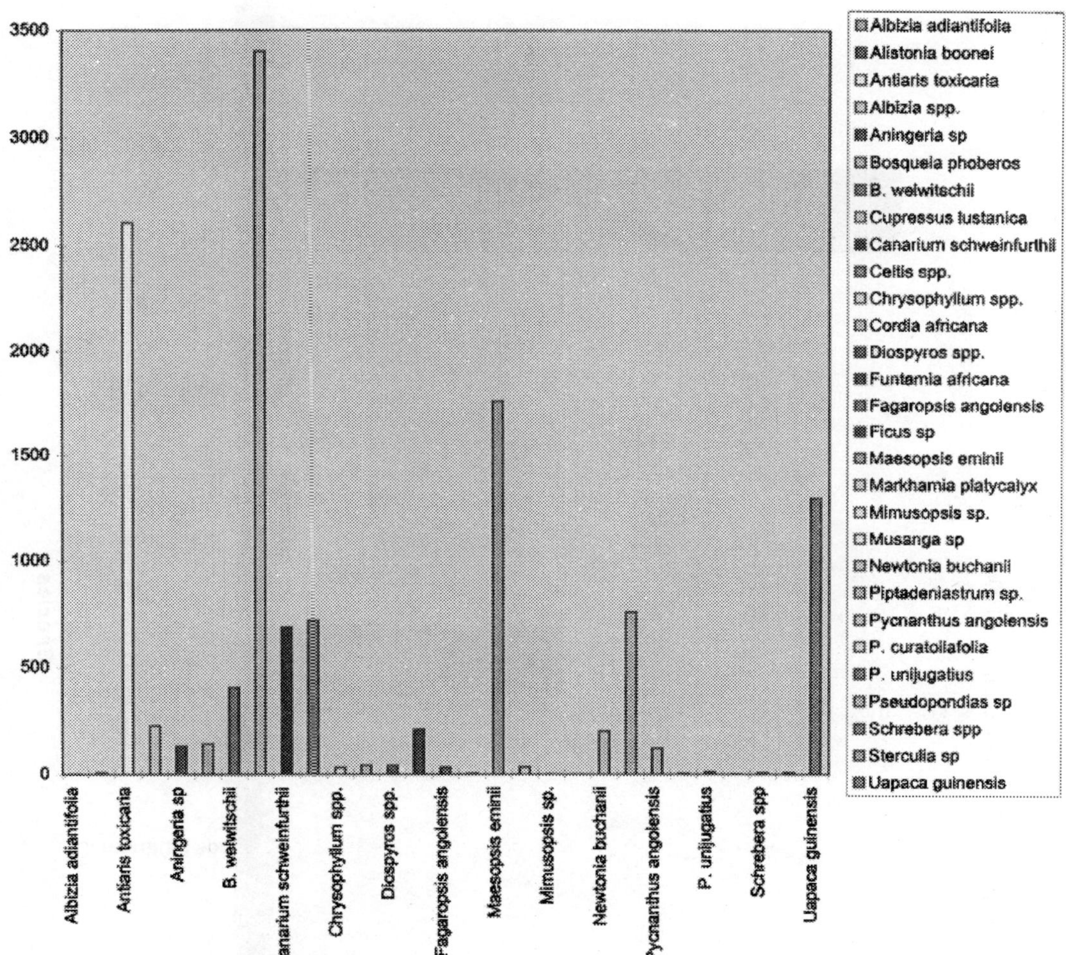

Source: Anon, 1997b

Figure 5.3: Class 3 Timber species marketed (June to December, 1996)

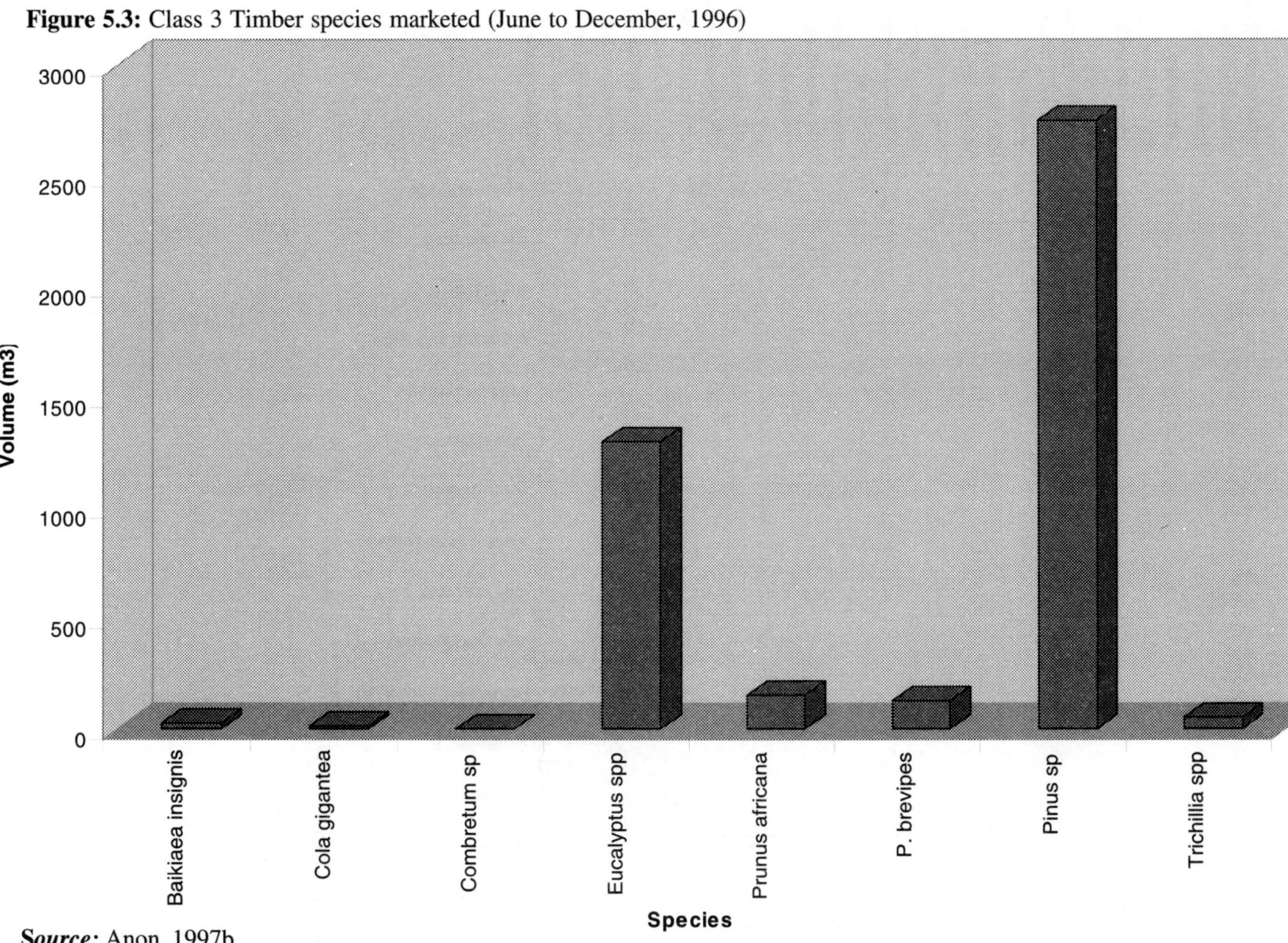

Source: Anon, 1997b

Hand sawing

Hand sawing is a method of converting logs into timber using hand saws. A platform is constructed and the log is rolled onto the platform. For very large logs which cannot be rolled onto a platform, a pit may be dug and the log rolled over the pit. One man stands in the pit or under the platform while the other stands on top of the log. The two men use a two-man hand saw to rip the log into planks or boards as shown in plate 5.5. Operation is hazardous since the log can fall onto the man in the pit. The use of platforms is more user friendly.

This trade resulted from the collapse of a large number of sawmills during the late 1970s and early 1980s. Pitsawyers supply over 90 % of all sawn timber from natural forests (Lubega, 1997). About 50 pitsawyers (employing a total of 200 men) were registered by the Forest Department in 1997 as shown in table 5.2 (Anon. 1997c). While it was envisaged that the Department would earn over 15,000,000/= (US $ 15,000) only about 30 % managed to pay the registration fees of 300,000/= (US $ 300) per year (Mpangire, per. com.). However, the response was better last year, when the registration fee was lower. It should be noted that there are a number of pitsawyers operating without licenses.

While Forest Department records indicate that less than 500 people are employed in the industry, it is estimated that over 6,000 men are involved in hand sawing (Anon. 1997b; Carvalho and Pickles, 1994). This may be caused by a number of unregistered pitsawyers or registered pitsawyers who use more than the recommended number of saws.

Pitsawyers operate both in natural forests and on public and private land. Pitsawyers have also played a key role in salvage operations especially in the aphid infested cypress plantations. Their operations are often wasteful as a result of converting trees into standard length logs and planks. In some cases, only the easiest or best part of the tree is sawn. This may be one log per tree! Large quantities of wood that could be converted to timber is often left to rot in the forest. During one survey it was established that over US $ 720 per ha is lost as a result of poor felling, bucking and sawing techniques (Norwegian Forestry Society, 1991). The Forest Department and Forest Research Institute are working with pitsawyers particularly in Budongo Forest to improve recovery in hand sawing.

Table 5.2: Registered pitsawyers per district (1997)

District	No. of registered pitsawyers
Hoima	23
Luwero	7
Mpigi	21
Mubende	8
Pallisa	2
Kabarole	17
Kalangala	10
Nebbi	3
Kamuli	12
Bundibugyo	4
Masaka	8
Kitgum	4
Arua	4
Bushenyi	2
Mbale	1
Lira	5
Gulu	1
Kibaale	1
Iganga	3
Mukono	12
Total	**151**

Source: Anon, 1997c

Logging and sawmilling

Natural forests

Prior to the 1970s there were 10 sawmills operating in the country based on natural forests. After the Asian owners left Uganda, most of these sawmills were operated as state enterprises. By mid 1980s only one mill was still operating.

Sawmilling in natural forest is limited to three sawmills although there are 10 companies registered with an estimated installed capacity (log input) of 88,500m³ (Carvalho and Pickles, 1994; Aluma and Ndimukulaga, 1996). Rwenzori Sawmill and Budongo Sawmill acquired equipment many years ago but the equipment was not installed. Four other companies imported equipment which is yet to be installed (Carvalho and Pickles, 1994). In 1990, only 6 mills were operating at 7.5 % capacity utilisation (Carvalho and Pickles, 1994).

None of the existing sawmills have sufficient suitable logging equipment. Sawmillers do their own logging under the supervision of the Forest Department. Sawmillers only pay for the merchantable volume of logs, often cut to a standard length of 4.2 m. As a result there is no log grading which discourages quality production. While Forest Department supervisors are expected to measure the logs and determine the volume using volume tables, it is not unusual for the sawmiller to deliver this information to the supervisor in the office! This practice has developed as a result of limited facilitation.

Often a large volume of unmerchantable logs and high stumps are left in the forest. Increased levy on first class timber species may help to reduce this practice. Proposed fees have been increased to Ushs. 72,000/= (US $ 72) from Ushs. 34,000/= (US $ 34) for class 1a which includes the valuable mahogany species (see appendix 5.1). In addition a number of class 2 species have been elevated to classes 1a and 1b (Anon. 1997d). These measures are designed to encourage efficient use of the resource. It is also envisaged that in the near future, the Forest Department will consider selling tree lengths rather than logs to encourage efficient use of the resource. This practice has started in Budongo forest, although implementation is still very difficult (Khaukha, per. comm.).

Poor logging techniques are used in most natural forests, with almost no directional felling causing tree hang-ups and in some cases standing trees may be broken. High stumps are often left behind, making logging even more difficult. Logs are skidded in the mud, often over long distances, to the landing site. Consequently, saw blades are blunted by the mud resulting in inaccurate dimensions and frequent break-downs.

Generally poor timber handling practices have continued to prevail at all hardwood sawmills. Timber is simply piled resulting in drying defects and rot. While efforts have been stepped up by the Forest Department to encourage better handling practices, the impact is yet to be realised.

Table 5.3 Sawmilling capacity in natural forests

Company	Area of operation	Estimated installed capacity input (m³)
Budongo sawmill	Budongo, Masindi	10,000
Amaply*	Budongo, Masindi	21,000
Bubwa sawmill	Budongo, Masindi	5,000
Kasenene Sawmill (Bukwa)	Budongo, Masindi	5,000
Rwenzori sawmill	Budongo, Masindi	7,500
Nkombe sawmill*	Kalinzu, Bushenyi	5,000
Nileply*	Mabira, Mukono	20,000
Kapkwata sawmill	Mt Elgon, Kapchorwa	5,000
Tesekererwa sawmill	Sesse Islands, Kalangala	5,000
Jinja Construction and Joinery	Public and private land	5,000
Total		**88,500**

Note: Installed capacity is based on log input.
* Sawmills are operating at very low capacity.
Source: Carvalho and Pickles, 1994; Aluma and Ndimukulaga, 1996.

Plantations
In anticipation of shortage of hardwood timber supply from the natural forests, the Forest Department established fast growing conifer plantations in the early 1940s. Over 14,000 ha of conifer plantations have been established in different parts of the country out of which 45% is mature (Kaumi, 1989). This is equivalent to over 27,000m³ of sawn wood per annum, although an estimated 10,000m³ was harvested in 1994 contributing between 9 and 13% to estimated demand (Natvig, 1996). Between June and December 1996 about 31% of all timber marketed was obtained from conifer plantations (Anon. 1997b).

The oldest sawmills in conifer plantations were established in the late 1960s and early 1970s in Kabale (Mafuga and Muko), Kabarole (Kyehara), Kapchorwa (Suam), and Nebbi (Lendu) Districts where the crop was ready for harvesting. By the beginning of the 1990s a large number of pine and cypress plantation forests were mature and ready for harvesting. The Forest Department encouraged investment in portable sawmills in order to exploit this resource. There are currently 51 sawmills operating in different parts of the country (Kyaroki, per. comm.)

The Forest Department operates three sawmills while the Forestry Research Institute operates one mill. These mills serve a dual purpose, that is, production of timber for sale, research and training.

Sawmillers obtain concessions from the Forest Department. Sawmillers do their own logging under the supervision of the Forest Department. In some plantations, logging takes place on hilly slopes where the gravity could facilitate log transport. However, minimal use of this advantage has been observed since most of the mobile sawmills are placed far from the logging sites and are often used as permanent mills. As a result, tractors and trucks are used to transport logs to the mill, often up to five km away from the logging site (Natvig, 1996). In some cases, downhill sites of the plantation have been harvested and sometimes replanted before the uphill sites thus making it impossible to effectively use the advantage of the slope in logging the areas further uphill. The logs are cut to 4.2m lengths and usually high stumps of 30 to 130 cm are left in the forest (Aluma and Ndimukulaga 1996).

Sawmillers only pay for the merchantable volume of logs, often cut to a standard length of 4.2m. As a result there is no log grading which discourages quality production. Often a large volume of unmerchantable logs are left in the forest as shown in Plate 5.6. Efforts are being made by the Forest Department to encourage low stump heights, log grading and variable log lengths in order to increase the recovery from the conifer plantations. Over 100 sawmill technicians have been trained in modern techniques of logging and sawmilling. In addition, there are plans to begin sale of tree lengths rather than logs so as to encourage better exploitation of the resource. In the long term, the Forest Department plans to sell standing volumes particularly for well managed plantations (Forest Department Technical Note, Forthcoming).

All sawmills installed in conifer plantations are operating below installed capacity as shown in Appendix 5.2. Besides the low capacity utilisation, there is substantial waste of significant proportions of the log volume. A study carried out by the Forestry Research Institute, established that recovery can be improved from 26% to 42% by lowering the stump height from 40cm to 5 cm, reducing minimum diameter from 12.5cm to 10 cm, log grading instead of bucking to standard length and re-sawing to obtain shorts from slabs.

Sawn timber, from plantations, public/private land and natural forests, is marketed through timber yards found in most urban areas throughout the country. However, large scale users especially the construction industry mainly order timber directly from the sawmills. Marketing of sawn timber provides a large number of jobs in urban areas.

Generally, the sawmilling industry provides employment for over 800 people and contributes 0.8% to the GDP (Ministry of Planning and Economic Development, 1997b). The sawmilling industry generates both skilled and semi-skilled employment opportunities for thousands of people in timber trade (including timber yards, carpentry workshops, construction industry and furniture marts).

Seasoning

Seasoning (drying) timber is almost non-existent in Uganda. Most sawmillers and pitsawyers do not stack timber orderly as shown in Plate 5.7. Such practices result in drying defects like warping, end splits and cupping. Such timber will often be rejected by customers and eventually rots or is used as firewood.

Most carpenters and construction companies use unseasoned timber. This not only reduces the longevity of the timber in use but also results in distortion of otherwise beautiful furniture and structures.

The few establishments (mainly Forest Department, Forestry Research Institute and a few sawmillers) that season timber use the air seasoning method. This involves stacking timber horizontally with stickers in between pieces of timber in a timber yard (Plate 5.7C) or vertically, leaning against wooden pillars (Plate 5.7D). Air seasoning is slow and is affected by changes in weather. It has been established from an on-going seasoning study conducted by the Forestry Research Institute that it takes between 9 and 21 weeks to air season different sizes of *Pinus caribea* (Anon., 1997e).

There are two establishments (Kapkwata and Roko Construction) that have installed seasoning kilns. These are the only companies in the country that use and sell seasoned timber. There is a kiln dryer at Budongo sawmill which has never been used (Natvig, 1996). Nileply Ltd has a biomass fired seasoning kiln used for drying plywood. The Forestry Research Institute plans to establish a small biomass fired seasoning kiln for research and demonstration purposes. It is anticipated that this technology will be disseminated to sawmills, timber yards and carpentry workshops.

Table 5.4: Sawmilling capacity in plantations (June 1997)

Type of Sawmill	Number	Unit Capacity	Total Installed Capacity
Locally Made	1	500	500
Locally Made	1	1,000	2,000
Kara	16	4,200	67,200
Lucas	9	1,500	12,000
Kamek	7	4,500	31,500
Wood Mizer	3	1,000	3,000
Wood Mizer	2	1,500	3,000
Mighty Mite	1	4,500	4,500
Mighty Mite	2	1,500	3,000
Forestor	1	2,500	2,500
Forestor	5	1,500	7,500
Laimet	1	4,000	4,000
Stenner	1	5,000	5,000
Bench Saw	1	1,500	1,500
Total	51	34,700	147,700

Note: Installed capacity refers to expected log input.
Source: Anon, 1997b.
See Appendix 5.2 for details of each sawmill.

Plate 5.1: Wood stacked for an Earth Kiln *(above)*

Plate 5.2: An Earth Kiln, half way through the production process *(below)*

Plate 5.3: Making charcoal from pine wood waste *(above)*

Plate 5.4: Charcoal packed in gunny bags for sale *(below)*

Plate 5.5: Hand sawing (*above*)

Plate 5.6A: Standard length (*below*)

Plate 5.6B: Logs abandoned in forest due to shorter length *(above)*

Plate 5.6C: Logs abandoned in forest due to shorter length and curve *(below)*

source: Aluma and Ndimukulaga (1996)

Plate 5.7A & B: Examples of badly stacked timber. Notice the mixture of sizes in B *(below)*

source: Ndimukulaga and Aluma, 1996

Plate 5.7C & D: Examples of stacking timber under shade by a sawmiller *(above)* and pitsawyer *(below)*

Source: Aluma and Ndimukulaga, 1996.

Plate 5.8 A & B: Demonstration of timber treatment by sap displacement *(above)* and dipping freshly sawn timber *(below)*

Source: Aluma and Ndimukulaga, 1996.

Carpentry and joinery

The carpentry and joinery industry is dominated by the private sector. However, the Forest Department has a carpentry workshop established many years ago. The equipment in use is obsolete, however, a team of well trained and experienced carpenters produce high quality furniture. Since mid 1993 the bulk of the furniture produced in the workshop has been dominated by pine timber. This has helped to encourage the use of pine timber for furniture.

There are about 15 large and medium scale furniture production companies in Uganda (Aluma and Ndimukulaga, 1996). These companies produce high quality furniture. Carpentry and joinery is a major source of self employment in urban and peri-urban areas. Most small scale furniture workshops produce low quality furniture with poor finishes. However, they supply the bulk of furniture used in the country.

Parquet

Parquet is used for flooring. There are machines at Rwenzori sawmill which have never been in operation (Natvig, 1996). Another machine was installed in Budongo Sawmill in early 1980s but has never been used (Natvig, 1996). Most of the parquet used in Uganda is imported. There are opportunities for investment in parquet production using the slabs and off-cuts from timber production.

Poles and posts

Transmission poles and fencing posts are mainly obtained from *Eucalyptus* spp. plantations. During the last 10 years there has been an increase in the use of palm trees (*Phoenix reclinata*) for fencing posts. The palms occurring naturally in swamps and along river banks are said to be durable. However, it has been established that untreated *Phoenix reclinata* is non-durable (Kamugisha, per. comm.).

The Forest Department has over 4,000 ha of *Eucalyptus* spp. plantations close to 14 urban centres. These plantations were established for provision of poles and posts. About 30% of the plantations are currently managed using an out-growers system (Byarugaba, 1997).

These plantations provide most of the building poles and transmission poles used in the construction industry and for electric and telephone grid expansion. Since the establishment of the out-growers scheme in 1993 the productivity of the plantations has improved from 335,000 tonnes in 1987 to 557,000 tonnes in 1997, valued at U. Shs. 8,896 million which includes both monetary (30%) and non-monetary (70%) values (Ministry of Planning and Economic Development, 1997b). In addition, a number of peri-urban households grow their food using the tangya system in these plantations particularly during the first year of establishment. Furthermore, peri-urban households obtain additional incomes from working in these plantations. The plantations also provide firewood although it is difficult to establish the quantities obtained.

A large number of indigenous species are major sources of building poles for domestic use. Most of these locally used species are resistant to termite attack, a serious menace in buildings.

Some termite resistant species include: *Acacia nilotica, Balenites aegyptica, Albizia zigya, Gliricidia sepium, Markhamia lutea, Melia azedarach, Oxytenanthera abyssinica, Terminalia brownii, Vitex doniana, Zizyphus abyssinica* (Katende, et al, 1995). Other species such as *Lophira alata* are resistant to fungi, termites and borers.

Peri-urban plantations and private woodlots supply all required poles and posts in the country. Companies that require large quantities generally purchase the poles in the forest. However, there are a number of private marketing outlets for poles, posts and treated posts close to timber yards and treatment plants. At times poles can be purchased from the road side close to the forest.

Phoenix reclinata, one of the species harvested for fencing posts, has been over exploited (Katende et al, 1995). Little or no effort has been placed in enrichment planting or plantations of this species. The species is also used for charcoal, roofing and basket making materials. The fruits are eaten and used to make wine.

Eucalyptus spp. grown in all peri-urban plantations is said to affect the soil-water balance, thus out competing food crops planted near eucalyptus stands. Although it has been accused of negative impact on soil fertility there is no scientific evidence to back this school of thought.

The Agroforestry programme of the Forestry Research Institute has established that other introduced species such as *Grevillia robusta, Casuarina* spp. and *Cedrella odoratta*, make good building poles, even when grown among crops (Wajja, per. comm.).

There is limited information on establishment of pole and fuelwood plantations from indigenous species. Studies on the potential of *Markhamia lutea* as a plantation species for provision of poles is under way by the Forest Management Programme of the Forestry Research Institute.

There are many households in Uganda who now obtain additional income from building poles. The short rotation period (4 years) for *Eucalyptus* spp. plantations has encouraged many individuals and organisations to plant trees as a cash crop. In addition, Uganda has indirectly contributed to reducing greenhouse gases in the atmosphere through establishing plantations which act as carbon sinks.

Pests of logs, posts, poles, sawn timber and furniture

The most important wood destroying insects belong to Orders Coleoptera (beetles) and Isoptera (termites). In the Orders Coleoptera, the Ambrosia beetles (Pinhole borers) are the most destructive. The adult beetles attack freshly felled logs whose moisture content is above 40%. They bore into logs, making narrow cylindrical galleries in the sapwood and also in the heartwood. The larvae feed on a fungus 'Ambrosia' growing on the walls of the tunnels. The tunnels become stained black, this often penetrating the wood for some distance upwards and downwards.

Freshly felled logs should be debarked and converted immediately to prevent infestation, while the sawn timber should be properly stacked and dried to moisture content levels below 40% (Gardner, 1957).

Powder post beetles

Infestation by these beetles is very common in timber yards and carpenter shops from where it may be carried unnoticed into houses in furniture, floor blocks and construction timber. The infestation can only be noticed when the new generation of adult beetles emerges through small circular holes 1mm by 1.5mm in diameter, accompanied by pushing out fine floury wood dust. Sapwood is the most preferred because of its starch content. Wood with moisture content of less than 8% is unfavourable for the development of the beetles (Gardner, 1957). Unfortunately, most air seasoned timber rarely attains such low moisture content.

Treatment of timber with preservatives such as tannalith and creosote prevents damage. Use of sapwood timber leads to serious damage and should be avoided. Painting, varnishing or oiling blocks the pores and prevents adult emergence.

Termites

Termites are serious pests of logs, posts and sawn timber. They bore into wood which is hollowed out leaving the outer shell intact. There are two biological groups:

(i)   Subterranean termites which are always connected with the soil. Wood even high up in the building may be attacked by this group since earth covered runways are built through the walls. Damage by this group of termites is most severe in north eastern Uganda, and in some dry areas in southern Uganda.

(ii)   Non-subterranean termites or dry wood termites: These have no connection with the soil. They live in isolated colonies especially in building timber on which they feed. Wood may be destroyed internally with the outside apparently intact but in reality, a thin shell.

The wood should be pressure impregnated with wood preservatives such as creosote and tannalith to protect them from damage by termites.

Preservation

Preservation is carried out to prevent insect and fungal attack. Wood for indoor use is not preserved. However, timber and poles used for construction purposes should be preserved. The common preservation treatment methods used include:
- brushing and spraying,
- dipping and soaking,
- diffusion treatments;
- vacuum pressure;
- sap displacement; and,
- hot and cold process.

Brushing and spraying
Organic solvents and oil borne preservatives are brushed or sprayed onto dry timber. Very little penetration or retention is achieved making this the least effective treatment method. This

method is used mainly for construction timber. Often spent oil or tar is used rather than recommended preservatives, thus exposing the timber to attack.

Dipping and soaking

Timber is totally immersed in a bath of preservative, thus wetting every surface. Organic solvents and water borne preservatives are used. Penetration and retention depends on immersion time and type of wood. Dipping is used for freshly sawn timber (Plate 5.8 B) while soaking can be used for dry timber. Dipping is mainly practised to prevent blue stain in pines and other perishable species using anti-blue stain chemical. However, this practice is limited since most customers do not reject stained timber (Natvig, 1996).

Diffusion treatment

Freshly sawn timber is immersed into a water borne preservative solution. The preservative must have a higher concentration than salts in the timber in order to be absorbed. The timber is left in moist, saturated conditions for about one month to ensure maximum penetration and distribution. This method is used for timber species which are resistant to traditional pressure impregnation.

Vacuum pressure treatment

External pressure is used to force liquid preservative into the wood. The method used in Uganda entails drawing an initial vacuum to pull air out of the wood cells, allowing the preservative to be pulled into the cylinder by the vacuum. The preservative is driven into the wood by applying pressure to the cylinder. Excess preservative is removed by applying a final vacuum. This is a highly effective and efficient treatment method but it is limited by the cost of equipment required. There are 13 pressure treatment plants in Uganda with a total capacity of over 109 m^3 as shown in Table 5.5 (Kityo and Plumptree, 1997). The Forest Research Institute plans to establish a pilot plant for research and demonstration. Vacuum pressure treatment is applied to timber and transmission poles. The utility and telecommunications companies have their own treatment plants although there are plans to privatise these plants.

Sap displacement

Sap displacement is applied to freshly felled fencing posts and poles (Plate 5.8A). It involves forcing water borne preservative into sapwood by pushing out sap. The posts under treatment are placed upright into a container with preservative. This method is used for fencing posts in most parts of the country .

Hot and cold process

The hot and cold tank process involves heating wood in creosote to a temperature of 80 - 90°C. During heating, air is expelled from the wood cells. On cooling the remaining air contracts, creating a partial vacuum which draws in surrounding preservative. This method is mainly used for fencing posts and transmission poles. It competes favourably with pressure processes. There are about 13 plants that use this method in the country (Kasirye, per. comm.).

Unfortunately, over 95% of the timber treated is wet (Aluma and Ndimukulaga, 1996). As a result, limited penetration is achieved. In the absence of timber treatment standards, and limited control exerted by the Forest Department, the Uganda Wood Preserves Association and the Forest Industries Development Association, the plants will continue to produce poorly treated timber.

Ten out of twelve preservation plants shown in Table 5.5 are currently operational. These provide employment for over 100 people (Kasirye, per. comm.). Preservation of timber and poles increases the longevity of wood in use thus reducing the demand for more timber and poles. While untreated poles and timber of some exotic tree species are perishable (0-1 year in use) to moderately durable (2-5 years in use) preservation of timber increases the years of wood in use to over 10 years. Proof of this is seen in the rate of replenishment of untreated fencing posts as compared to well treated transmission poles. Information on the durability of indigenous species is limited.

Preservatives on the market today are chlorinated hydrocarbons and arsenic based products (Kityo and Plumptree, 1997). Most plants using these products do not have proper disposal mechanisms. As a result some chemicals may be deposited in soil and water bodies. Workers in treatment plants should wear protective clothings. It has been established that long exposure to preservatives can result in certain hazards (Kityo and Plumptree, 1997). However, no new preservatives have been developed therefore existing chemicals will continue to play a leading role in wood preservation. It is imperative to note that wood which may come into contact with food or drinking water should not be treated. In addition, treated timber should not be burnt in open fires or used for cooking.

Table 5.5: Preservation plant capacity

Plant	Capacity (m³)	Location
East African Sawmills (2 units)	10	Kampala
Kapkwata sawmills	10	Kampala
MKM Corporation	12	Kampala
BM Technical Services	6	Mbarara
NIFRA	4	Mbarara
Budongo Sawmills Ltd	4	Masindi
Kagera sawmill	7	Kampala
Uganda Post and Telecommunications	7	Kampala
Nileply	9	Jinja
UEB (creosote)	15	Kampala
UEB (creosote)	15	Tororo
Kilembe Mines Ltd	n/a	Kasese
Moon Enterprises Ltd	n/a	Kampala
Forest Department-mobile unit (sold)	n/a	Kampala
Total capacity	>109	

Source: Aluma and Ndimukulaga, 1996; Kityo and Plumptree, 1997

Panel products

Plywood, blockboard, hardboard and softboard

Panel products are used in the furniture and construction industry. Limited quanties of plywood is produced by Nileply which mainly exploits *Antiaris toxicaria* from Mabira Forest and West Mengo Forest. Peeler logs are obtained mainly in private forest. However, pine logs have also been used for production of plywood. Nileply also produces limited quantities of blockboard. There is no local production of soft board or hard board.

The bulk of panel products used in Uganda are imported. In 1996, importation of panel products was estimated at US $ 1,245,000 (MPED, 1997b).

There is potential for investment in the panel products industry. Existing sawmills in Uganda can expand their activities to include production of panel products. This would further improve the utilisation of the forest resources, provide additional employment to Ugandans and more revenue to the Government.

Pulp and paper

Pulp production is not considered to be viable. Uganda imports paper and pulp for paper production. Importation of paper and pulp in 1996 was estimated at U.Shs. 17,508,000 and UShs. 26,000,000 respectively (MPED, 1997b).

Other wood products

Other wood products include hand tools, carvings, musical instruments, sporting goods, utensils, weapons (bows and arrows; gun butts, batons), wheels/spokes. Producing these goods provides part time employment for a large number of people in rural areas. This also improves the utilisation of the forest resource since mainly branches of trees felled or pollarded are used.

Non-wood forest products

Rattan/Cane and other Fibres

The use of rattan or cane for furniture making is a growing industry in Uganda. However, there is little information on quantities produced and level of resource exploitation. Between January and June 1997 over UShs. 5,000,000/=(US $ 5,000) was collected as revenue from non-wood forest products (Anon., 1997b).

Handicrafts such as baskets, mats, and other decorations are also made from different species. The most commonly used species are *Ficus natalensis* and *Ficus ovata* for bark cloth, a traditional fabric used in central Uganda. Other species include:

Borassus aetiopum an indigenous palm tree is used to make baskets, mats and brooms;

Ensete ventricosum (Wild Banana) is used for rope and sack making;

Fibres of *Grewia mollis* are used to make string for general use and construction of granaries;

Phoenix reclinata is use for making mats and baskets;

Piliostigma thonningii is used to make ropes;

Raphia farinifera is used to make decorations, furniture, ropes and baskets; and,

Smilax anceps is used for furniture, baskets and fish traps.

There are many other species whose fibres are used for rope making and thatch. Exploitation of fibres from the forest ecosystem is still limited to subsistence levels. However, the handicrafts industry has continued to grow and is expected to thrive as the tourism industry improves. A large number of rural households obtain additional incomes from handicraft sales.

Extractives

Extractives such as latexes, oils, gums, resins are collected from live trees for industrial applications. One of the major foreign exchange earners in Uganda in the early 20th century was high quality rubber from *Funtamia elastica* (LTS, 1995). It is no longer a major export product. Latex from *Ficus platyphylla* and *F. elastica* are used locally for tool handles and musical instruments (Katende, et al, 1995).

Gum from *Acacia senegal* (gum arabic) is used in the food textile, cosmetic, pharmaceutical, lithography and printing industry. Despite the high demand for export and domestic use, there is limited exploitation of gum arabic in Uganda. Gum arabic was once an important source of income for households in Bokoro county, Karamoja region where the Forest Department was a major dealer. Over 3,900 kg of gum was collected from natural exudates in 1973/74 (Kityo, 1991). A study conducted by the Forest Research Centre established that there is a reasonable stock of *Acacia senegal* in Karamoja region producing high quality gum that satisfies international (JECFA) specifications. Over 4,000 kg were tapped in an area of 2 km radius (Kityo, 1991). The study revealed that there is a potential for 15,000 to 20,000 kg of gum collected from natural exudates (Kityo, 1991). Tapping would yield even more gum.

Other species which provide gums used in the above mentioned industries include *Acacia seyal, Ficus platyphylla* and *Anacardium occidentale*. However, gum from these species is inferior to gum arabic. No studies have been carried out in Uganda on the extent of the resource and quality of gum from these species.

Resins are tapped mainly from mature *Pinus caribea* stands. Resin is currently exported to Kenya and is used for production of turpentine and rosin. Rosin is used in manufacture of adhesives, paper sizing agents, printing inks, surface coatings, insulating materials for electronic industry, synthetic rubber, chewing gum, soap and detergents. Thirty workers can tap about 20,000 kg of resin in one month, providing about 200,000/= (US $ 200) in revenue. However, resin tapping is limited to a few plantations and is not carried out throughout the year. Considering that there are over 5,000 ha of mature plantations, resin tapping needs to be encouraged and export markets sought.

Tannin is an extract from tree species used in the leather industry. Commercial exploitation of tannin has mainly been from *Acacia mearnsii* an introduced tree species. However, other species such as *Acacia nilotica, Sizygium* spp., *Prosopsis africana, Parkia africana, Trema orientalis* etc. also provide tannin (Katende, et al, 1995). Commercial exploitation in Uganda is limited and there is hardly any information on the quality of tannin produced by the different species.

Different colour dyes are obtained from a large number of tree species. There is no commercial exploitation. Dyes are exploited locally for handicrafts.

Oils from different tree species are extracted for the pharmaceutical, cosmetic and soap industry. Common species from which oils are extracted include *Eucalyptus globulus, Eucalyptus ciriodora* (source of citronella), *Vocanga thouarsii, Butyrospermum paradoxum, Azadirachta indica, Ximennia* spp., *Ricinodendron heudelotii* etc (Katende et al, 1995). However, commercial exploitation is limited to a few species.

Medicines

A large number of tree species are used for medicines. Common ailments for which medicines are extracted locally from trees include:

Malaria   -   bark of *Azadirachta indica* (Neem), and leaves of *Tetradenia riphria*.

Cough   -   roots of *Acacia hockii*, leaves of *Catha edulis, Ecuridaca l ongipenduclata, Tetradenia riphria*; fruits of *Stereospermun kunthianum*; and, bark of *Callistemon citrinus, Tetrapleura tetraptera* and *Zanthoxylum gilletii* (Katende et al, 1995).

Other ailments for which medicine can be obtained include whooping cough, asthma, boils, mental illness, epilepsy, stomach ache, chest pain, diarrhoea, dysentery, venereal diseases, vomiting, worms, leprosy, sore throat, fresh wounds, toothache, ulcers, impotence, hypertension, mumps, meningitis, and many more (Katende et al, 1995). Additional information on herbal medicines can be obtained from 'First National Plant Genetic Resources Workshop: Conservation and Utilisation' 1996, in a paper by Kakooko and Kerwagi entitled 'Medicinal Plants in Uganda' and 'Medicinal Plants of East Africa' 1993 by Kokwaro.

Toxins and pesticides

Toxins extracted from *Acokanthera schimperia, Adenium obesum* and *Euphorbia tirucalli* are mainly used as arrow poison and for fish poisoning (Katende, et al, 1995). *Balanites aegyptica* is used to kill snails and water fleas, the hosts of bilharzia and guinea worm parasites. Powder from the leaves of *Azadirachta indica* (Neem) are used to fumigate against seed borers in grain storage (Katende, et al, 1995). Neem is also said to repel termites. *Phytolacca dodecandra* fruits are used to kill mosquito larvae in ponds. *Melia azedarach* is used to control insects attack on vegetables. A large number of species are used for different purposes although there is hardly any commercial exploitation.

Food

Food from the forest is often the least discussed topic. Different tree species provide fruits and nuts for both humans and wild animals. Bamboo shoots make tasty stew in the Mount Elgon area. Some species provide vegetables too. Many tree species provide fodder for wild animals.

Eco-tourism

Eco-tourism is one of the latest sources of income in the forestry sector. Eco-tourism was started in 1993 with support from the European Union under the Nature Conservation Project of the Forest Department. There are currently five eco-tourism sites in Budongo (2), Mabira, Mpanga and Kashoya-Kitoma forest reserves. This is part of a joint-collaborative forest management system with communities who live close to the forest and it has proved to be an important source of revenue for the communities and the Forest Department. In addition, three forest reserves (Semuliki, Mt Elgon and Bwindi) are now managed as national parks by the Uganda Wildlife Authority.

Institutional role

The Forest Department

The Forest Department is the major forest estate owner. It is a government department. The Department gives forest users concessions or licences to exploit the resource and supervises exploitation activities. In addition, the department plays a key role in collecting revenue from the forest estate.

Training institutions

The major forestry training institutions are the Forestry Department, Makerere University and the Forestry College, Nyabyeya, a section within the Forest Department. The University offers degree and postgraduate courses in Forestry while the College offers diploma and certificate courses; and refresher courses. The Forest Department's Utilisation Section also plays a key role in refresher courses and on-job training. Other technical institutions offer training in carpentry and joinery.

Over the last three years, the training institutions have seriously addressed training requirements for the forest products sub-sector, to address the problem of inefficient utilisation of forest resources.

Research institutions

The Forestry Research Institute and The Forestry Department, Makerere University carry out research in forestry. The Forest Products Programme under the Forestry Research Institute conducts research in Forest Products. During the last 3 years research activities have concentrated on efficient harvesting and sawmilling, timber properties and seasoning. The programme plans to address many other issues, some of which have been raised in this chapter.

The Forest Industries Development Association
The Forest Industries Development Association is one of the major stakeholders in the forestry sector. It is an association which brings together private companies which operate different forest related industries.

Other major stakeholders

The Uganda Wildlife Authority is the second largest forest owner. Their involvement in forest products is limited to conservation and minor harvesting of different products by the communities living close to the forest. However, they have converted natural forest into major tourist attractions thus exploiting one of the non-tangible forest products.

The Uganda Wood Preservers Association is also one of the associations which brings together major stakeholders in wood preservation. Unfortunately, it has been relatively inactive.

Other major stakeholders are charcoaliers and wood fuels traders. However, they are not organised into large associations. There are a few associations at district levels.

The silent but most important stakeholder group are those people who live close to the forest estate and have for many years used and conserved this resource. Most of the information on non-wood forest products and use of indigenous species in this text actually comes from their wide knowledge and experience. Since these stakeholders are unable to write, those who extract the information often get the credit!

References

Aluma, J.W.R and Ndimukulaga, J.P. 1996. Report on Wood Industry Sector in Uganda and Recommendations for Improvement. Small Enterprise Development Company & Uganda Manufacturers Association (UMA) March 1996.

Annon. 1992. Technical Report: National Biomass Study Phase I. The National Biomass Study, Forest Department.

——— 1994a. Schedule I Part A - Timber Royalty, Forest Produce and Licence Forest Department Records.

———1994b. Forestry Development Programme, Forest Department.

———1995 Study of Woody Biomass Derived Energy supplies in Uganda. ESD report, Uganda Forest Department.

——— 1997 a Periurban Plantations, Reports from Different Districts, Forest Department.

——— 1997 b. Timber monitoring Database. Forest Department.

——— 1997 c. Pitsawyers file. Forest Department

——— 1997 d. Draft Revised Forest fees. Forest Department

——— 1997 e. Progress reports, Forestry Research Institute.

Boiling Point, 1991. More Charcoal by Traditional Methods. *Boiling Point* No. 24, April,

Byarugaba, D. N. 1997. Current status of Peri-urban Plantation in Uganda. Forest Department, Uganda.

Carvalho, J. and Pickles, G. 1994. Forest Industries Rehabilitation and Development Plan. FRP G1824-U. Ministry of Natural Resources, Republic of Uganda.

Gardner, J.C.M. 1957. Insects Injurious to Timber in East Africa. Proceedings of the 14th meeting of the East African Timber Advisory Board, Moshi, Tanzania.

Hibajene, S. H. 1994. Assessment of Earth Kiln Charcoal Production Technology. Republic of Zambia, Ministry of Energy and Water Development, Dept of Energy and Stockholm Environment Institute. Energy and Environment Series No. 39.

Hong, J.C. 1995. Indoor Air Pollution and its Challenge in China: Response Options for the Household Energy Sector. FWD Working Paper.

Karekezi, S.; Turyareeba, P.J. 1994. Biomass Energy Initiatives: Experiences in Eastern and Southern Africa. Paper presented at the Solar Energy Society of South Africa, Regional Forum, South Africa.

Katende, A.B.; Birine, A; Tengnas, B. 1995. Useful Trees and Shrubs of Uganda: Identification, Propagation and Management for Agricultural and Pastoral Communities. Published by Regional Soil Conservation Unit,

Kaumi, S.Y. 1989. Report on Softwood Plantations in Uganda. Forest Department

Khaukha, S. 1997. Timber Species Utilisation in Budongo. In: Woodsman, Issue No. 4

Kityo, P.W. 1991. "Natural Gums and Resins (Uganda)". Technical Report R-86-0239. Forest Research Centre, Forest Department, June 1991. 25p.

——————, and Plumptree, R.A. 1997. The Uganda Timber Users' Handbook: A Guide to Better Timber Use. OFI, Commonwealth Secretariat/Government of Uganda.

Kyaroki, A (Anon) Brief of the current management of Forests in Uganda, 1997. Forestry Department, Uganda.

LTS International Limited, 1995. Evaluation of the EDF-Funded National Forest Management and Conservation Project. A component of the World Bank Forestry Rehabilitation Project, Uganda. 1988-1992. Volume 2.

Lubega, C. 1997. Timber Report Scary. *The New Vision,* Friday September 19.

Mabonga-Mwisaka, J. 1972. The Mark V Portable Steel Kiln. Uganda Forest Department Technical Note 197/72, A revision of Technical Note 164/70.

Ministry of Natural resources, 1994; The State of the Environment Report for Uganda.

MONR, 1994; The State of the Environment Report for Kampala Uganda.

MPED, 1997a. Background to the Budget 1997/98 (and medium term development strategy). The Republic of Uganda. Ministry of Planning and Economic Development, Kampala.

MPED, 1997b "Statistical Abstracts" The Republic of Uganda, Ministry of Planning and Economic Development, June 1997.

Natvig, T. 1996. A Study of the Wood Industry Sector in Uganda. Republic of Uganda, Ministry of Industry and Trade, Uganda Manufacturers Association Consultancy and Information Services(UMACIS), Small Enterprise Development Company (SEDCO). With support from Swedish International Development Cooperation Agency (SIDA) .

Norwegian Forestry Society, 1991. Report on Combined Forestry Training at Nyabyeya Forestry College. Second Draft, November.

Turyareeba, P.J. 1990. The Performance of Four Woodburning Cookstoves and the Relationship Between Wood Properties and the Performance of Two Woodburning Cookstoves. A Thesis for M. Phil., Aberdeen University, UK.

Turyareeba, P.J., 1991. "Improved Charcoal Production in Uganda, A Case Study." Paper presented at the Regional Training Course on Biomass Energy held in Nairobi and Nakuru, Kenya, June 1991.S

UNDP/ESMAP, 1996. Uganda Energy Assessment UNDP/ESMAP, Joint UNDP/World Bank Energy Sector Management Assistance Programme. Power Development, Efficiency and Household fuels Division, Industry and Energy Department. Draft Document.

WHO, 1992. Indoor Air Pollution from Biomass Fuels.

Classification of timber tree species

Fee Group I   **Trade/local name**

Group 1 a

Milicia excelsa	Muvule, Muvule, Iroko
Entandrophragma angolense	Mukusu, Gedu, Nohar
Entandrophragma cylindricum	Muyovu, Sapele
Entandrophragma utile	Muyovu, Kikula, Muhungura
	Utile, Budongo heavy mahogany, Mufumbi
Khaya anthoteca	Munyama, African mahogany
Khaya grandifolia	Munyama, Tido
Khaya senegalensis	En, Tido
Lovoa trichilioides	Nkoba, Mukusu
Lovoa swynnertonii	Nabulagala, Nkoba, Mukyenyi
Olea welwitschii	Elgon olive, Elgon teak
Olea africana	Elgon olive, Elgon teak, Lilondo
Olea hochstetten	Lilondo
Podocarpus gracilior	Podo, Musenene, Sabtet
Podocarpus mianjianus	-do -
Podocarpus usambarensis	-do -
Albizia conaria	Mugavu, Musisa
Cordia millenii	Mukebu, Mukomeni, Mutumba
Cedrella odorata	Cedrella

Group 1 b

Fagara lepneurii }	Ntareyeirungu, muramankobe,
Fagara macrophylla }	East African Satin wood, Munyenye,
Fagara mildbraedii }	Omushanga, Mutabembwa, Kitumbe.
Afzelia africana	Afzelia, Beyo, Meli, Azza
Guarea cedrata	Scented Guarea
Hagenia abyssinica	Omugeshi, Leo, Sigurura, Naturu
Holoptelea grandis	Mumuli, Mutawale, Butungu, Mutaa
Mitragyna rubrostipulata }	Nzingu, Abura
Mitragyna stipulosa }	-do-
Octea usambarensis	East African comphorwood, Omwiha
Cupressus lustanica	Cypress

Fee Group II	**Trade/local name**
Pinus spp. (all species)	Pine
Albizia ferruginea	Nongo, Mulong, Omushebeya
Albizia grandibracteata	Bulera, Ajuara Adzimel
Albizia gummifera	Owak, abat, bedo, ebatata, swessu, oulera, muchole, segavu
Albizia zygia	red nongo, bulera, mushebeya, nkwasi
Carapa grandiflora	Uganda crabwood, crabnut, Mutanga, Mujojo
Fagaropsis angolensis	Muyinja, Mafa, Kabwegi
Ficalhoa laurifolia	Mwumanga, Omuvumanga
Maesopsis eminii	Musizi, Muhongera, Omusinde, Muguruka
Markharnia platycalyx	Musambya, Omusavu, Aborigo, Erniti
Newtonia buchananii	Muchenche, Lupewere, Mugaye
Piptadeniastrum africanum	Dahoma, Agooin, Mpewere, Mugaye
Prunus africana	Red stinkwood, Ngwabuzito, Ntasesa
Symponia globulifera	Musaali, Muyanja, Musisi, Munimba
Trichilia dregeana	Sekoba
Alangium chinese	Mukoko, Omikofe, Curonon
Albizia adianthifolia	Bulera, Komosovi, Mushesebyea
Albizia glaberrima	White Nongo
Allophylus abyssinicus }	Biomwa, Gulindi, Kirindi
Akkiohtlus macrobotrys }	-do-
Alstonia boonei	Mujwa, Alstonia
Aningeria adolfi-friederic	Sosi, Mwiruni
Aningeria altissima	Osan, Mutoke
Antiaris toxicaria	Kirundu, False iroko, Antiaris
Balanites wilsoniana	Naligwalimu, Desert date
Canarium schweinfurthii	Mwafu, Incense Tree
Cassia mannii	Cassia
Cathorrnion altissimum	Muchere
Celtis adolfi-frederici	Akanisa, Musisa, Nyabinunka, Mastet
Celtis africana	Stinkwood, Mulindu, Namumuke, Musia
Celtis durandii	-do-
Celtis integrifolia	-do-
Celtis rnibraedii	Lufugo, Mukomakoma
Celtis zenkeri	-do-
Celtis wightii	
Chrysophyllum albidum	White star apple, Mululu, Nkalate, Muhambulya
C.goruggosanum }	Muhambulya, Monkey star, Munyamba
C.mnerense }	-do-
C. pentagonocarpum }	
C. perpulchrum }	
Cordia africana	Mukebu, Mutumba, Mujungangoma, Mugengere
Cynometra alexandri	Iron, Muhirnbi

Fee group II	**Trade/local name**
Diospyros abyssinica	Nkinga, Mpojwa, Muhoko, Cheptua
D. mespiliformis	West African Ebony, Cumu
Dombeya bashawei	Nkokwa, Mukorabo
D. goetzenii	Gabaluwa, Mukole, Borowa
D. mukole	Mukole
Drypetes spp.	Mushabara
Ekebergia canpensis	Musalamu mal, Mufumba, Bumet
E. senegalensis	Mufumba
Erythrina abyssinica	Muyirikiti, Kakiri dua Muko, Olua
E. excelsa	Mubajangabo, Mulungula
E. mildbraedii	-do-
Erythrophleum sauveolens	Mumara, Earamor, Odiodi
Faurea saligna	Mororia, Mulenjere, Mukuka, Moyoku
Funtamia Africana	Nkago, Munyamagosi, Munyumatuga
F elastica	Nkago, Musanda
Ilex mitis	Mwandanda, Munyambasi, Bwiso, Sigora
Macaranga kilimandscharica }	Mudwess, Muburashasha, Muhunga,
Macaranga lancifolia }	Muhoti, Omusheshe, Omurara
M.monandra }	-do-
M. schweinfurthii }	-do-
Mildbraediodendron excelsum	Muyati, Maburere, Bambo
Morinda lucida	Mubajansaki, Mukiringi, Omusinganjovu
Morus lactea	Mukonge, Nyakatoma
Musanga cecropoides	Kaliba, Namuguro, Kigere, Kikmbu
Myrianthus holstii	Mugunga,Kiruhura, Echuvu, Kibende
Myrica salicifolia	Mukembo, Omudirindi, Omujeje
Parinari excelsa	Mubura, Ebwa, Namulambo, Omushambo
Parkia filicoidea	joge, Mujojo, Omusheshe
Polyscias fulva	Setala, Mungu, Omungu, Omurungi
Premna angolensis	Mutala, Nkulwa, Banyumunkiro
Pseudospoindias microcarpa	Muziru, Bagambanimpyata
Pterygota mildbraedii	Omwifa, Mukoko, Ndaula, Mwiha
Pycanthus angolensis	Lunaba, Mukoyoto, Muno, Ongomo
Schrebera alata }	Ndeta, Nawaluma
S. arborea }	-do-
Spathodea campanulata	Nandi flame, Kitabakazi, Munyera, Opal
Sterculia dawei	Kitokwe, Musandasanda, Omuhongo
S. setigera	-do-
Syzyguim guineense	Kulungisanvu, Omukonda, Omugete
Tetrapleura tetraptera	Munyegenye, Kikangabarimi
Warburgia ugandensis	Mwiha, Musizambazi, Balwagira
Acacia spp.	-
Juniperus procera	African pencil cedar, Ntorkya, Torokio

Fee Group III	**Trade/local name**
Baikiaea spp. (all)	Nkobakoba
Balasamocitrus damei	-
Balthasaria schiebenii	Omukali
Cassipourea spp. (all)	Pill erwood, Aganiya, Kobwo
Cistanthera kabingaensis	Mbaka
Cola giggantee	Mutumbwe, Omujugangoma
Danielle olivera	African capaiba Balsum, Bitole
Klainedoxa gabonensis	Mukuzanyama, mutuufu
Manilkara aborata	Nkunya
Mimusopsis spp. (all)	Musandasanda, Musali
Ochuo holstii	Lokotono, Siteli
Olinia usamberensis	Nerekia
Patcystela brecipes	Nkalate
Pappea capensis	Mulemambazi
Spandinnthus presussii	Mimbiri, Mutambuzi
Staudatia stipulata	-
Strombosia scheffleri	Munyankore, Omuhika
Tapura fischeri	Kazunganjuki, Barerewa
Xylopia spp. (all)	-
All others	-

INDEX